THE FUTURE
OF
SCIENCE

THE FUTURE

OF

SCIENCE

By Herbert

Morris Goran

SPARTAN BOOKS
New York • Washington

Copyright © 1971 by Spartan Books.
All rights reserved. This book or parts thereof may
not be reproduced in any form without permission from
the publisher.
Library of Congress Catalog Card Number 71–133106
International Standard Book Number 0–87671–704–0
Printed in the United States of America.
Sole distributor in Great Britain, the British Common-
wealth, and the Continent of Europe:
       Macmillan & Co. Ltd.
       4 Little Essex Street
       London, W. C. 2

For My Family

Books by Morris Goran

Introduction to the Physical Sciences. Glencoe, Ill.: Free Press, 1959.

Experimental Chemistry for Boys. New York: John F. Rider Publisher, 1961.

Experimental Biology for Boys. New York: John F. Rider Publisher, 1961.

Experimental Astronautics. Indianapolis: Howard W. Sams & Bobbs-Merrill, 1967.

Experimental Earth Sciences. Indianapolis: Howard W. Sams & Bobbs-Merrill, 1967.

The Core of Physical Science. Chicago: Cimarron Publishers, 1967.

Experimental Chemistry. London: Lutterworth, 1967.

The Story of Fritz Haber. Norman: University of Oklahoma Press, 1967.

Biologia Experimental. Barcelona: Ramon Sopena, S.A., 1967.

CONTENTS

INTRODUCTION

This volume is addressed to several audiences. First are students involved with science, whether they intend to make it their future livelihood or are merely enrolled in courses in general education science, science and public policy, or history of science. Those majoring in areas other than science are as concerned with the nature of the scientific enterprise as are the chemists, physicists, and biologists. The interest may appear to be monetary, in that public money is the chief supporter of research today, but the student of the social studies and humanities usually demands more than the financial audit. For those majoring in science, engineering, medical, dental, and public health fields, the science of science is a necessary rather than peripheral study.

This book is also addressed to scientists as members of a profession that seeks to assess its present prospects and to envision its future. Old, established trades and professions have the benefit of strong, wealthy organizations to do much of this investigation; for example, the American Medical Association has a great deal of help in evaluating the present and future status of its membership. At times the older organizations are even aware of an obligation to the general public, and they make their findings widely known.

The science professions are not yet in a position to have a unified approach. The American Association for the Advancement of Science and the National Academy of Sciences is each a broad umbrella for scientists of all kinds in the United States, but neither is the equivalent of the American Bar Association nor of the American Institute of Accountants. Scientists are not too interested in belonging to broad groups. They give their loyalties to their specialty organizations such as the American

1

Chemical Society or the American Physical Society. But the fact that they are not unified is not an indication that they are not inbred with the historic background of their work. If anything, they are more knowledgeable about their history than are attorneys, accountants, or physicians about theirs.

Between April 13 and 19, 1969, twenty-nine scientists invited by the American and British Associations for the Advancement of Science met in Boulder, Colorado, to assess the future of science, but no record was published save a two-page report in *Science,* the organ of the American Association for the Advancement of Science, on August 8, 1969. The report indicated that there was "more stock-taking than crystal-gazing, more thoughtful meditation about the present than brainstorming about the future."

This book seeks to look into the future for science and scientists, using past records as a basis for prognosis. The first chapter focuses on three problems of modern science—communication in all its aspects, financing and allocation of resources for research, and education. The remainder of the book describes events of contemporary and past science that have a bearing on these issues, particularly communication and education. The whole array is presented in the hope that intelligent men and women in all walks of life will be concerned and interested.

1
GROWING PAINS

As recently as 1940, it was socialistic heresy in the United States to propose government support of scientific research; those who advocated national laboratories were often called communists; the National Bureau of Standards was said to be ample intrusion of government in science. World War II and its aftermath changed this perspective. Science now receives munificent government funds. Science is now an Establishment and is Big Science; it has taken its place along with other modern giants —business, labor, government, education, and religion. The child of the Renaissance nurtured by patrons and dilettantes has grown to manhood. Science is now big in terms of research workers and budget. It has been reported that 90 percent of all scientists that have ever lived are alive today. The amount of money spent for research in the United States is about 3 percent of the gross national product.

All giant institutions foster a bureaucracy, and science is no exception. The wheels of science are turned mainly by people active in the National Academy of Sciences, government circles, and such select professional organizations as the American Chemical Society. One group that has great influence is the Committee on Science and Public Policy of the National Academy of Sciences, COSPUP. Since its inception in January 1962, this committee of fifteen men has influenced national science policy more than any other assembly of scientists. By June 1966, at the end of close to five years of existence, the Committee had published reports on the growth of world population, on basic research and national goals, on federal support for university research, and on investigations of the needs of astronomy, computer science, physics, and chemistry. COSPUP's members—a chairman and a representative from each of the fourteen sections of the National Academy of Sciences—meet approximately one weekend every month and serve on a rotating basis for terms of about three years. With one or two exceptions, their reports have been prepared by a panel of scientists and reviewed by the Committee as a whole.

The wheels of science have been in need of lubrication and careful examination ever since the spurt of activity during World War II. Public money has supplied most of the lubricating oil, but careful examination has been rare. Three major areas in which such an examination would reveal problems are communication, criteria for support of research, and science education.

Communication is a magic word to those seeking quick and general solutions to complex problems; it rivals education in proposed applicability to the difficulties of mankind. Misunderstanding, lack of comprehension, and marital relations are said to be helped by communication. In all these cases and others where communication is prescribed, the intent is to differentiate between effective and ineffective relaying of information.

Science and technology have developed very powerful media for communication, but none have helped them very much in scientific meetings. During the 1920's and 1930's, virtually all the attendants at the meetings of the American Physical Society recognized each other; the informal conversation which ensued was often educational. But after World War II, the membership grew and the American Physical Society was faced with the dilemma of large gatherings. Other professional societies in science have the same problem. When the members of the American Chemical Society met for the 94th time, in September 1937, there were 469 reports; at the 132nd meeting, in September 1957, there were 1,408 reports. Thirty-eight hundred chemists attended the Houston, Texas, meeting in March 1970, and an editorial in *Chemical and Engineering News* following the meeting pleaded: "Smaller meetings—fewer technical papers—more face-to-face discussion—closed circuit TV—absence of big-city congestion and frustrations—all will help." In 1900 the American Association for the Advancement of Science had a membership of 1,925; in 1914 the number had risen to 8,325; now there are more than 100,000 members.

Scientific meetings last from three to five days, and during this time the attendant generally pushes his way from one crowded room to another. Cloakroom discussions are hardly cultivated when a scientist must rush from one hotel to a second, often blocks away, within a short time period. Meeting others in the same general field of investigation does not seem to be fostered when social mixers are held in rooms where there is scarcely elbow room available. Many papers are read concurrently, and the scientist must make an important choice: whether to spend ten minutes to listen to X or to Y. When many such decisions are necessary in any one meeting, the attendant may decide on alternative educational activity—taking in the sights of the city.

Small gatherings of scientists are by invitation only, and the young, ambitious research worker is generally not present. Moreover, when specialists communicate only with each other, they deny themselves the benefit of interaction with scientists outside their specialty, and thus they deny themselves knowledge of other areas. The situation is described in Nobel-prize winner Harold Urey's confession to a *New Yorker* reporter during the summer of 1969. Urey's oldest daughter is a nuclear physicist and his son is a biologist. Urey said: "Do you know, I don't understand a thing either of them talks about."

If the research scientist has a finding to announce and avoids the presentation of a paper at a meeting, he can submit a manuscript to one of the learned journals serving his field. If a panel of his peers—the editorial board and the referees—decides that his contribution is not worth publishing, he can always seek another journal outlet. There are several new ones each year. The number of scientific and technical journals in the world is between 25,000 and 50,000, excluding the thousands of technical government reports and house organs which may publish scientific work; one exact count during the 1960's found the number to be 35,300. A 1960 U.S. Senate Committee report claimed that the science and technology collections in the Library of Congress have doubled approximately every twenty years during the past century.

Journals seeking to abstract this flourishing knowledge must become more expensive; in 1966 a subscription to *Chemical Abstracts* cost $1,200, a sizeable percentage of the total periodical budget of a science library. Even with necessary funds, scientists seeking to keep up with their own narrow field of investigation have an increasing amount of difficulty.

In 1966, the National Academy of Sciences set up SATCOM, a committee on scientific and technical communication, to explore and help solve problems in such areas as journal publication, indexing, abstracting, and cataloguing, in order to insure that knowledge flows more easily to the seeker after knowledge. A chief proposal among the fifty-five recommendations in their 1969 report is the establishment of a Joint Commission on Scientific and Technical Communication responsible to the councils of the National Academy of Sciences and the National Academy of Engineering.

In 1966, a promising experiment in communication was terminated after five years of growth. Members of information exchange groups (IEG) in very specialized fields of inquiry, mainly in the biomedical sciences, communicated with each other very rapidly. Financed by the National Institutes of Health, the IEG distributed any kind of communication from comments, protests, and review articles to experimental results and argumentation. Nearly eight hundred members were in the first group, and at the demise of the experiment there were six other groups, all smaller than the first, costing the National Institutes of Health about $400,000 per year. The quality of the communications was said to be no worse than that found in regular journals, where a wait of six months or more is usually required for publication, the format prevents much controversy, and articles are reviewed by referees before acceptance for publication. The first group, whose subject was electron transfer and oxidative phosphorylation, had more than three hundred members outside the United States, and a mainland Chinese applicant was in the process of being accepted when the program stopped. It was widely accepted that pressure from the editors of U.S. and British biochemistry journals was the main cause for the end of the system.

In 1970, some members of the American Psychological Association were disturbed by an expensive plan, implemented to establish a similar system in psychology. Some psychologists feared control by a central office staff, and that the quality of psychological literature would deteriorate. The concept involved a computerized procedure for distributing unedited manuscripts rapidly. Designed with the help of almost $1 million from the National Science Foundation, the procedure attempted to overcome the swell of information in psychology. The number of so-called core journals in the field more than doubled from 112 to 226 between 1960 and 1970; the number of psychological abstracts more than doubled from 8,532 in 1960 to 19,586 in 1968.

The fragmentation of the scientist has occurred as a result of the knowledge explosion. In contrast to the universalist of earlier days, the present-day scientist must be a specialist. Physicists are now experts in weak interactions, solid state, or cryogenics; the biologist is known as the student of the Krebs cycle, behavior of the Australian rabbit, or DNA synthesis. The scientist generalist is looked upon with disdain.

Specialization has resulted in scientists unable to communicate effectively with intelligent non-scientists. Sir Charles Percy Snow has raised this issue in an illuminating way in his book *The Two Cultures and the Scientific Revolution*. Snow and others assert that there is little communication, and sometimes hostility, between the humanist and the scientist. Science and its good companion technology, goes the argument, have lessened poverty, hunger, and disease, and have become the hope of the underdeveloped countries; the literary intellectuals are often unaware of these contributions and indeed may be antagonistic to the scientific revolution even when it brings desirable social change. Snow claims the following are either ignorant of science or are anti-science: Henry James, T. S. Eliot, Ezra Pound, William Butler Yeats, D. H. Lawrence, Virginia Woolf, Franz Kafka, and William Faulkner. Several years ago, writer Mark Van Doren said in one of his humanities classes, at Columbia University, "I hate science." Literary critic Joseph Wood Krutch wrote in June 1969 that science for science's sake, that is, for pursuit of knowledge without thought to how it may be applied, is no longer a tenable aim.

Lesser-known members of the literati are also in the camp.

Christopher Williams, reviewing Jean-Luc Godard's film *Alphaville* in the English journal *New Society,* wrote: "No scientist, no computer expert should delude himself for a moment; basically we're skeptical, we don't trust you, and we're not grateful . . . you dine too often at the boss's table; you've never stood up successfully against the military. And then we're crowded together in sprawling soulless cities, having things *done to us*. We can't describe exactly what the things are, but we do know that you haven't come to our aid yet." French critic Jacques Ellul has written that scientists are "sorcerers who are totally blind to the meaning of human adventure" and that their system of thought is bringing about "a dictatorship of test tubes rather than hobnailed boots." American critic Paul Goodman claims that many young people find science and technology to be anti-life. The July 31, 1969, issue of the *New York Review of Books* had an article entitled "Technology: Opiate of the Intellectual."

A similar treatment to science was given by others in the non-English speaking world. Spanish philosopher Miguel Unamuno took the anti-science approach in his 1913 book *The Tragic Sense of Life;* Spanish philosopher Ortega y Gasset did likewise in his *The Revolt of the Masses,* published in 1930. In the 1960's, the Swiss playwright Friedrich Duerrenmatt's *The Physicists* was a brutal attack on the callousness of scientists engaged in a race for the power to destroy the earth.

In 1967, Dr. E. Orowan wrote in a letter to *Science* (v. 157, p. 874): "The vast majority of the Earth's population regard science and technology as an increasingly mortal threat to their lives. They feel themselves powerless at the mercy of a few, as if they were on the operating table in the hands, not of healers, but of irresponsible playboys driven by curiosity, if not by the desire for prestige and promotion. It would be a good thing if scientists realized that they are dancing on a powder keg."

The attitude of the scientists has been engendered by the needs of science. To be a specialist demands concentration and effort through college and graduate-study years. The aspiring scientists of today have little or no time to devote to the values to be derived from art and literature. If their home life cultivates the taste for art and literature, they are fortunate, because modern curricula do not. Snow, however, is impressed with the efforts made in some American universities to give science students more humanities courses.

The narrow specialization may be defeating for science in the end. A specialty can become obsolete; it can be thoroughly searched and researched. New specialties arise and only those scientists with resiliency and broad knowledge seem to be able to move away from their stratified training into new, richer realms of study. The biologist who went to school in the 1930's has none of the deep training in physics and chemistry that is necessary for the lively and growing field of molecular genetics. Such a biologist and other scientists must join ranks with the many professionals who need an occasional retooling of their education.

Scientists not only have a problem in communicating with humanists and other scientists, but they also have the gnawing need to communicate with nonscientists in general. Popular science has a long history of support—but not from scientists. At first, popular science meant writing in the vernacular instead of Latin. At times such accomplishments as Galileo's were as much literary as scientific, and they influenced great numbers of people. During Newton's time Voltaire was one of the most successful popularizers. His compatriot Bernard de Fontenelle was also an accomplished science writer; his writings went through six editions during his lifetime and six more by 1825. During the nineteenth century some first-rank scientists in Europe gave popular lectures and some wrote for the public. One of the best traditions then introduced was the Christmas Lectures at the Royal Institution in London. Some twentieth-century scientists have continued to try to talk to the public, but most research workers look upon the activity as unrewarding. Fortunately, the profession of science writer has drawn its membership from journalists, teachers, and scientists. Improvements are necessary, for example, the creation of a *Scientific American* for less scientifically sophisticated readers. They too need to know how their tax money is being used for scientific research. In the last analysis, the public is the supporter of science.

During the early stages of the great spurt of scientific research in the United States, Congressmen responsible for allocating and approving the expenditure of public funds were more or less uncritical in their acceptance of the needs of science. This lack of diligent questioning continued on until the time of the first Soviet success in space in 1957; the United States was supposed to be in the scientific forefront, and legislators and others eagerly sought to support United States activity in science. The result was that federally financed scientific work in space, atomic energy, and defense absorbed roughly two thirds of the total supply of physical scientists and engineers. During World War II such similar concentration of research workers proved effective for the development of the atomic bomb, the proximity fuse, and radar.

Science today has a problem acquiring funds. Public money must be voted by legislators, and scientists will find it difficult to compete with other burgeoning demands. The Vietnam war, for example, brought a slowdown in federally sponsored scientific work. The reduction in federal support for scientific research was met with tears, indignation, threats, and protest. Requests sometimes escalated into demands, and prestigious groups and individuals cited dire consequences. The New York Academy of Science in its report "The Crisis Facing American Science" called for an annual 15 percent increase in federal funds to avoid the catastrophes forecast.

Before the twentieth century, scientific investigation was much lower in cost; it was no strain on the pocketbook of the patron. Some naturalists had enough money to finance their own research. Or they could hold undemanding jobs, by which they could support themselves and their research. Another procedure was to go with a tin cup in hand to some philanthropist known to be friendly to science. These methods of financing are still valid for the few who can be so accommodated.

Industrial laboratories grew in size and number during the first half of the twentieth century. They have become a new, important source for research and development funds. With or without tax provisions designed to encourage industrial research, the private sector is a receptive, albeit restrictive, financier for research; its money goes largely for development.

Government funds for scientific research, which account for the bulk of it in any country today, is a late twentieth-century phenomenon, despite its surfacing now and then in prior years. Aristotle had room and board with the Macedonian hierarchy but not for scientific research; his biology work was done later. Leonardo da Vinci had a job with a dictator, but Leonardo's scientific work was done on his own time. Tycho Brahe received "tuns of gold" for his island observatory and then lived well under Rudolf II of Bohemia. William Gilbert did receive some money from Queen Elizabeth, but he was also her physician.

When states needed the help of scientists during times of war, science was supported. At the end of the eighteenth century, chemist Fourcroy wrote about "the importance of the geometrical and physical sciences and the necessity of harvesting their fruits" during war. Napoleon was helpful to the men who at times gathered at Arcueil because he thought he needed them for his exploits. Meteorology was supported by France and England after disasters to their ships in storms during war.

During World War I the Allies were slow to realize the benefits of science. When the American Chemical Society offered its services, they were declined on the grounds that the War Department already had a chemist. On the other hand, the Germans had a well-organized Chemical Warfare Service that was an important factor in their staying power.

Everyone knows that wars were not responsible for the ideas of Galileo, Newton, Darwin, Einstein, and all of our celebrated scientists. However, war funds have supported a large percentage of lesser-known scientists of our time. It would be reasonable for such men and women to find financial support now and in the future by latching on to wars other than the hot and cold ones between the Western and collectivist powers. Two such wars are the crusade against pollution in all its aspects, from purification of air and water to the restoration of good living in the cities, and the battle to thwart cancer. Both require many a pure science effort. Another war is the war against military struggles of all kinds; which emphasizes the peace-loving and international character of science. Until organisms from outer space threaten our survival, these are the only wars that can again benefit scientific research.

In 1966, Ronald F. Probstein and several colleagues in the

Fluid Mechanics Laboratory in the Department of Mechanical Engineering at the Massachusetts Institute of Technology decided to reduce their involvement in military and space research and become more active in socially oriented research. The laboratory then had six professors, twenty graduate students, and an annual budget of about $300,000. In 1969, there were nine professors, thirty graduate students, and a budget of about $600,-000. About 35 percent of the 1969 budget was defense-oriented whereas in 1966 it was virtually all in that category. Not one of the professionals at the laboratory in 1966 was an expert in air and water pollution, biomedical fluid mechanics, or the desalination of water. Nonetheless, government agencies supported the research.

Many specialized science groups have published plans for more tax money. The Interagency Committee on Oceanography of the Federal Council on Science and Technology asked for $2.2 billion for the ten-year period 1963–73. The President's Commission on Heart Disease, Cancer, and Stroke recommended a five-year program to cost more than $1.6 billion. A 1965 report of chemists calls for the expansion of basic research by 20 percent for three or four years and 15 percent for several years thereafter. Meteorologists recommended increasing the appropriation for their science to at least $30 million a year by 1970. The 1966 National Academy of Sciences study "The Plant Sciences" calls for a total of $1.5 billion in support from federal agencies until 1974. The physics group asked for a 250 percent rise from the 1963 level of federal support. Even if the total federal research and development budget increases at an average rate of about 20 percent per year, as it did between 1948 and 1964, there will still not be enough money for these and other similar demands.

A few scientists have challenged these vast proposals because they feel that the allocation of public funds for research cannot grow continuously. One, Alvin Weinberg, director of the Oak Ridge National Laboratory, has proposed that relative social value of the research serve as the chief criterion for support. Late in 1967, William D. Carey of the Bureau of the Budget amplified such a scheme at a Johns Hopkins Applied Physics Laboratory colloquium. His paper "Social Merit Matrix," which he presented as a "beginning to stimuate reaction and discussion," suggested that social merit be measured in terms of the contribution of the research to economic, cultural, and political values.

The 1970 report of the President's Task Force on Science Policy suggested one-tenth of one percent of the gross national product as a reasonable level of support for the National Science Foundation. The report recommended that the NSF be responsible for about one-third of federally-funded basic and applied research and that the Office of Science and Technology be the focal point in the executive branch of government for establishing priorities for scientific research programs rather than having them set by default by the Bureau of the Budget.

If pure science is to be financed and judged by applied science needs, researchers would be wise to consider pressing for international financial support. Many other countries are accenting applied work. General industrial work is being done at Britain's Harwell Atomic Energy Establishment. The country's Science Research Council report for 1967–68 stated: "The Council knows that more scientists could profitably be engaged in fundamental research, but the number available is limited and it is essential to the country's prosperity that a greater proportion, especially of the most able, should be applying their skills to tackling the country's industrial problems." At the 1968 general meeting of the U.S.S.R. Academy of Sciences, a proposal was made that the Academy do certain projects in applied science.

The situation is similar in the United States. During the 1956–64 period, federal support, mainly military, for basic research grew 29 percent a year while in 1964–69, its growth rate was only 9 percent a year. During the summer of 1966, President Lyndon Johnson said: "A great deal of basic research has been done. I have been participating in the appropriations for years in the field. But I think the time has now come to zero in on the targets by trying to get our knowledge fully applied." It could be that pure science has become too expensive for the wealthiest nation. The so-called Mansfield amendment, adopted late in 1969 as Section 203 of the Military Procurement Authorization Act of 1970, prohibits the Defense Department from financing "any research project or study unless such project or study has a direct or apparent relationship to a specific military function or operation." A member of the House Subcommittee on Science Research and Development said early in 1970: "Section 203 sets an example that is contagious and it's going to spread very quickly to other agencies." The president of the National Academy of Sciences also said that the idea "has begun to rub off on other agencies."

International support for pure science could possibly nurture scientific talent throughout the world, giving an opportunity to many new disenfranchised by the accident of place of birth. International pure science may be another antidote for the divisiveness among people; when many problems demand worldwide solutions, the unity rather than the diversity of man needs to be featured. Independent support on a global basis, perhaps as a function of UNESCO, would free pure science from the taint of the alleged evils of technology, and the subject's own beauty and worth could be better appreciated. Once financial support is assured, however, the problem of priorities will vex pure science.

The assessment of research projects will have little guidance from prior experience. Scientific research has been supported for such reasons as curiosity, national security, philanthropy, salesmanship, friendship, and hero worship. Seldom, if at all, has a scientific project been supported to help the underdeveloped. Yet today, underdeveloped universities and underdeveloped regions of the United States press for more funds. Those not receiving a sizeable share of federal contracts have been outspoken in calling for a greater equalization of funds. In 1964, the National Science Foundation began to consider seriously awarding millions of dollars in grants to universities capable of developing into top research centers, and several were awarded. Later, in 1967, they had a more modest means, in economic terms, for improving college and university science departments—the College and University Science Improvement Program (COSIP).

The universities are beginning to be concerned with a systematic study of allocation for research and its associated problems. Two pioneers in the area overseas are the science studies unit at the University of Edinburgh, Scotland, and the research policy program at Sweden's Lund University. Late in 1966, Columbia University inaugurated an Institute for the Study of Science in Human Affairs. Purdue University has recently established a division to study science and public policy, as have Harvard and Cornell Universities.

The universities will not be the spokesmen for the kinds of scientists who need financial support and who have been very productive, according to the records of the history of science. The dedicated amateur and the professional who had no organizational affiliation were responsible for great feats during the past centuries. If a very recent case is typical, the chance for the encouragement of such people is very small.

The case in point is that of Lieutenant William Fox of the New York City Police Department. Fox is a veteran police officer who earned a Ph.D. in chemistry at Columbia University in sparetime study. He had received his undergraduate education at the City College of New York during the time of economic depression in the United States. When he was graduated in 1935, jobs of any kind were scarce, and winning an appointment to the police force was a plum for economic security. He stayed with this job while seniority and other benefits accrued. In 1949, he took a leave of absence for a year to work in the Bureau of Mines petroleum experiment station at Bartlesville, Oklahoma. He continued to do scientific research in his basement laboratory, following lines of investigation that interested him, not necessarily those that were in fashion or done by large organizations. He published his results in such first-rate journals as the *Review of Scientific Instruments* in 1950 and the *Journal of Physical Chemistry* in 1959.

Fox applied to federal agencies for monetary help. In 1953, when the National Science Foundation was in its infancy and not yet a bureaucracy, he received $225 from them to attend a three-week conference on instruments at New York University. Following this success, however, all his applications for funds were fruitless. Thinking that federal agencies would be more sympathetic to organizations than to individuals, he began to apply under the name of the Oakland Research Associates, with William Fox as the principal investigator. The Office of Naval

Research told him that "the statutory authority of the Office of Naval Research permits the making of grants only to educational institutions and certain other nonprofit organizations and only for the purpose of basic research." They did not support his application for funds. In 1961, the National Aeronautics and Space Administration invited him to submit a proposal, but then they turned it down. In 1963, the National Science Foundation refused to award him one of the six travel grants it gave to a Brussels conference in his specialty. Fox attended anyway, using his own funds.

The apparatus of federal support for scientific research is not attuned to help the Foxes in this country. Yet it is in the equivalent of today's basement and garage laboratories that much fundamental discovery and invention has been made. The greatest technical and scientific achievements of mankind—the control of fire, the discovery of the wheel, the domestication of plants and animals, and the use of simple machines—were the accomplishments of amateurs. One of the founders of physical anthropology was the French attorney Edward Lartet; the first ethologist in the present sense of the word was the American lawyer Lewis Henry Morgan. Even in applied science there are professionals who were once amateurs. In the 1930's, Richard S. Perkin was a New York stockbroker and amateur astronomer. He helped found the Perkin-Elmer Corporation in 1938, and this organization was a major supplier of optical equipment for the United States government during World War II.

The federal agencies also make a mistake in their attitude toward the vast numbers of amateur scientists who have little formal education. There is no program of monetary support for amateur astronomers and geologists, two of the many kinds of amateurs who often band together for social and other benefits. The last planet to be discovered—Pluto in 1930—was found by an amateur, Clyde Tombaugh, while the professionals at the largest observatory had missed it. Another case is that of Douglas R. Emlong of Oregon who spent years gathering the world's largest collection of marine fossils, along the wet and windy beaches of Oregon. In 1967, the Smithsonian Institution and the University of Oregon argued over the rights to this collection while 25-year-old Emlong became an art education student at Lewis and Clark College in Portland.

Even if the amateurs never engage in research, their activities should be promoted because of their educational value. Work with amateur groups can often lead a youngster to a career in scientific research.

Relative neglect of science education is a third major problem of expanding science. This would seem to be a strange charge in the light of the National Science Foundation, as well as other foundations, financing course- and curriculum-improvement projects in science. In September 1966, the National Science Foundation issued a 118-page outline of the many activities it was supporting in this area, in addition to its in-service institutes, grants, and fellowships for teachers.

A European panel studying American science education came to the conclusion in 1963 that it was a growing American problem. The Organization for Economic Cooperation and Development (OECD) is composed of eighteen member nations; it appointed a distinguished three-man panel to study technical manpower needs and technical education in the United States. Panel members were British physicist Sir John Cockcroft, British sociologist A. H. Halsey, and Swedish economist Ingvar Svennilsen. They concluded that in the near future the United States will have a shortage of people with graduate degrees, although there will be no lack of men and women with undergraduate training. They recommended direct federal help to education for the purpose of leading more students to the Ph.D. degree. They cited that a much smaller percentage of U.S. college students study science compared to their counterparts in Britain and the U.S.S.R. About 20 percent of the undergraduates in America are science and technology majors, while the figure is 45 percent in Britain and 57 percent in the U.S.S.R. Besides the smaller percentage, the United States is handicapped too, reported the panel, by the inferior quality of teaching in high school and college. The federal government, they said, is responsible for this predicament because it funnels too much money into research and not enough into education; the smaller liberal arts colleges have been "starved" for the benefit of a few big institutions because energy has been diverted from teaching into research. The panel suggested that it might be wise to reduce expenditures on aerospace projects to make possible the release of more science and engineering talent for other fields.

The more firm recommendation was for direct federal support of teaching.

The American science community for the most part ignored this report as they did a 1968 OECD larger publication on American science policy. Comments in *Science* scored this last document as being based on a rapid survey rather than on exhaustive scholarship.

No one seriously challenges the contention that the federal program of supporting research has adversely affected the teaching of science. College and university faculty members who were already attuned to placing research at the pinnacle of their totem pole of experience have been more or less encouraged further in this value system by federal activity. The researcher has a wide variety of grants, contracts, fellowships with accompanying emoluments of travel aid, apparatus purchases, and miscellaneous overhead expenses. The teacher has none of these financial and prestige incentives to excel in teaching. The researcher is rewarded with reduced numbers of classes, leaves of absence, and publicity. The teacher has, sometimes, the rewarding thanks of a small number of students. The researcher has the aura of sainthood and the title of scientist. The teacher is associated with the routine, ordinary, and undramatic. There is a vast amount of money available for researchers, and the art of obtaining this money has been cultivated by some who are known as scientific entrepreneurs and grant-acquiring experts; they know how to acquire the money but not necessarily how to follow through with acceptable research. Teachers do not have a chance to develop this nefarious talent; they can locate no similar fountain of money.

Almost every college and university can attest to how their science teachers have been beckoned with federal research contracts to the detriment of their teaching function. Of course, an institution seeking an image as a research center, or overhead money (the administrative funds supplied by the government), welcomes the grants and encourages its staff to participate. In 1964, one small college had at least ten cases wherein research awards had been given without the college administration even being consulted; the awards had come from six different agencies, one federal and five nonfederal. Representative Devine of

Ohio, supporting the reduction in federal spending for scientific research, said in the fall of 1969: "University professors have become independent contractors living on federal grants and have given up the art of teaching and being professors."

The 1964 annual report of the Carnegie Foundation for the Advancement of Teaching places some blame for the situation on the faculty member. Their claim is that for these teachers, "students are just impediments in the headlong search for more and better grants, fatter fees, higher salaries, higher rank. Needless to say, such faculty members do not provide the healthiest models for graduate students thinking of teaching as a career."

Graduate education in the sciences needs to have more emphasis on internships for teaching. Data obtained by Alfred J. Lotka and Derek J. de Solla Price at Yale University has shown that 60 percent of research scientists publish only one research paper early in their career and then cease to publish at all. Perhaps these are the same ones who are dissatisfied as teachers, unable or unwilling to fulfill a graduate-education-inspired goal of research. Teaching as a career should be presented to them in a more favorable way.

A college or university serious about improving teaching will need federal support to help educate its staff. Many college teachers are deficient in the number of science courses studied, let alone in knowledge of contemporary science. An analysis of the applicants for a 1963 summer conference in physics revealed an amazing lack of education in physics in college physics teachers. The National Science Foundation in cooperation with the Commission on College Physics set up three conferences on mechanics at the University of Colorado, Dartmouth College, and Southern Methodist University. Three hundred and forty-nine persons applied and eighty-two participants were chosen. This approximate four-to-one ratio of applicants to places available is average for the NSF Institutes for College Teachers. The ratio is higher in the high school teacher program, and it is about fifty to one in the case of elementary school institutes. The number of semester hours of physics studied by each of the applicants was listed. The average number was twenty; the mode was in the range from eleven to fifteen.

The typical undergraduate curriculum in physics requires

thirty to forty semester hours of study in physics. Therefore the American college student majoring in physics may be taught by someone very poorly prepared, someone without enough physics courses to win an undergraduate degree in physics. Of 349 college physics teachers, 347 did not have enough training in college physics to be a graduating senior in physics at a typical American college.

If eighteen semester hours of college study in a subject be defined as a minimum for high school teaching, then a 1963 report of the National Science Foundation reports that 66 percent of high school classes in physics in the United States are taught by inadequately trained teachers. The condition has since been aggravated because in 1966 the American Institute of Physics reported that the number of undergraduate physics majors had decreased since 1962. More than one third of the first-year graduate students in physics in 1965–66 dropped out of physics before getting a degree. The proportion of the male freshman college class of September 1968 choosing physics as a major concentration was .82 percent, down from .90 percent in 1967, from 1.27 percent in 1963, and from 1.78 percent in 1960. A report in the September 1969 issue of the Commission on College Physics *Newsletter* claimed that only two schools out of seventeen hundred in the United States produce more than ten high school physics teachers per year, and only ten schools produce at least five physics teachers per year. The largest, most respected physics departments produce only one high school physics teacher every five years.

The conditions are comparable in other areas of science education. In the spring of 1965, a questionnaire was sent to 8,463 teachers listed as earth science teachers in the U.S. Registry of Junior and Senior High School Science and Mathematics Teaching Personnel, compiled by the National Science Teachers Association. The 3,224 replies received were analyzed and showed that 53 percent of the teachers had never taken a course in astronomy, 62 percent had not studied meteorology formally, 80 percent had never enrolled in an oceanography course, and 50 percent had never been formal students of physical geography. The data showed that 94 percent had received six or less credits in astronomy, 96 percent had six or less credits in meteorology,

99 percent had six or less credits in oceanography, and 85 percent had six or less credits in physical geography. Less than a third of the teachers had sufficient college credits for a typical teacher's major, twenty semester hours of credits, in any of the earth sciences.

The chances now are better that the situation will improve and science teachers will be more properly prepared because there is increasing concern for such social problems as education among young people and scientists. On December 4, 5, and 6, 1969, a "Conference on Graduate Preparation for Teaching— The Missing Component" was sponsored by the Commission on College Physics and held at the University of Washington. The students who took part showed a keen desire to change procedures at universities so that the teaching function might flourish.

SELECTED REFERENCES

Brooks, H. "Physics and the Polity," *Science*, 160 (April 26, 1968), 396–400.
DeReuck, A., Goldsmith, M., and Knight, J., Eds. *Ciba Foundation Symposium, Decision Making in National Science Policy*. Boston: Little, Brown, 1968.
Gilman, W. D. *Science, U.S.A.* New York: Viking, 1965.
Goldsmith, M., and Mackay, A. *The Science of Science*. Baltimore: Penguin, 1966.
Goran, M. "Future Scientists Need More Encouragement," *Chemistry*, 39 (March, 1966), 15–21.
———. "The Literati Revolt Against Science," *Philosophy of Science*, 7 (July, 1940), 379–84.
Greenberg, D. S. *The Politics of Pure Science*. New York: New American Library, 1968.
Hornig, D. F. "United States Science Policy: Its Health and Future Direction," *Science*, 163 (February 7, 1969), 523ff
Klaw, S. *The New Brahmins: Scientific Life in America*. New York: Morrow, 1968.
Lapp, R. E. *The New Priesthood: The Scientific Elite and the Uses of Power*. New York: Harper & Row, 1965.
Moravcsik, M. J. "Private and Public Communication in Physics," *Physics Today*, 18 (March, 1965), 23–26.
Moore, John A. *Science for Society: A Bibilography*. Washington: The

Commission on Science Education, American Association for the Advancement of Science, 1970.

Orlans, H. "Some Current Problems of Government Science Policy," *Science,* 149 (July 2, 1965), 37–40.

Orlans, H., Ed. *Science Policy and the University.* Washington: Brookings Institution, 1968.

Waddington, C. H. "A British Perspective on American Science Policy," *Science,* 160 (April 5, 1968), 46–48.

Walsh, J. "Congress Subcommittee Surveys Effects of Federally Supported Research on Higher Education," *Science,* 149 (July 2, 1965), 42–44.

Weinberg, A. "Criteria for Scientific Choice," *Minerva* (Winter, 1963), 159–71.

————. "In Defense of Science," *Science,* 167 (January 9, 1970), 141–45.

Wolfle, D. "The Support of Science in the U.S.," *Scientific American,* 213 (July, 1965), 19–25.

2
MISTAKES WITH MONEY

A commonplace belief in the United States is that "money talks." If this slogan is believed to offer the solution of the problems of communication, allocation of resources, and education in science, how will scientists fare?

John Mills was an early twentieth-century American engineer and popularizer of science. At one time he claimed that the order of interest of scientists was thus: things, ideas, people, and, lastly, money. Since scientists and engineers are involved with laboratory apparatus as well as with the materials of the world, the first item seems logical; science is a place where ideas are tested, and so the second is in line with accepted thought; laboratories and observatories are not as crowded as are theaters and auditoriums, and by the laws of chance alone scientists would be dealing with people relatively less frequently than would other professionals; and money—it is accepted folklore that scientists are not concerned with money.

Visitors to chemical laboratories may often see a quotation posted that more or less substantiates the disinterest of scientists in money matters. J. J. Becher in 1669 wrote: "Ye chymists are a strange class of mortals impelled by an almost insane impulse to seek their pleasure amid smoke and vapour, soot and flame, poisons and poverty. Yet among these evils, I seem to live so sweetly that I may die if I would change places with a Persian king." Also cited are the words of philosopher and astronomer Eudoxus, who wrote about two thousand years ago: "Willingly would I burn to death like Phaethon were this the prize for reaching the sun and learning its shape, size, and substance."

The folklore about scientists and money is supported by such expressions as that of the nineteenth-century naturalist Louis Agassiz, who said, "I have no time to make money." Joseph Henry, director of the Smithsonian Institution for thirty-two years, refused many positions "especially because it might be supposed I am influenced by pecuniary reasons." Michael Faraday belonged to a religious sect that believed the accumulation of money to be immoral. Albert Einstein asked for a tiny sum when he was first employed at the Institute for Advanced Study in Princeton, New Jersey.

The history of science, however, reveals a different picture. Among those who obtained ample funds from sympathetic monarchs were biologist Aristotle and astronomer Tycho Brahe. The latter was supported in magnificent style on the island of Hveen for about twenty years. In 1588, he reported to the new government, shortly after the death of his royal patron, King Frederick II, that because of his scientific work he was in debt two and one-half times the amount of money he had been getting per year. When Galileo Galilei was a struggling scientist, he had as many as twenty students renting quarters in his house. Early in his career he devised a hydrostatic balance which sold well enough for him to hire workers for its manufacture. Isaac Newton was a very successful investor and died a wealthy man. Later, German-born William Herschel capitalized on his ability to design and build telescopes; he earned a great deal of money from their sale. Herschel married a wealthy woman and could then pursue science in elegance. Other pioneers also married well, or else they had wealthy patrons who supported their work.

During the nineteenth century, some scientists did not hesitate to gather up money. William Thomson, Lord Kelvin, had many patents and became wealthy. J. J. Thomson was an astute investor; the size of his estate at his death surprised his intimate friends.

During the early twentieth century, Walther Nernst developed a useful infrared source and promptly marketed it for a million marks. His contemporary Fritz Haber, developer of the first successful nitrogen fixation process, insisted upon a flat sum for every pound of material sold; he reasoned that the price would go down as the process was perfected, and a flat sum rather than a percentage royalty would insure a stable, large income. However, he did not foresee the disastrous consequences of German post-World War I inflation; but the company he had bargained with luckily made a new financial arrangement with him.

After World War II, scientists imbued with the spirit of free enterprise established manufacturing and consulting organizations; in a few instances, these organizations have grown to be giants of industry. Scientists have been welcomed into the councils of finance, having been elected members of the boards of directors of profit-making organizations. In some cases, scientists and engineers have rivaled the high income of physicians and surgeons.

More money has come to scientists from prizes and medals. For example, the Fermi award was authorized in 1954 to be given by the Atomic Energy Commission "for any specially meritorious contribution to the development, use, or control of atomic energy." The law set no amount for the prize. Enrico Fermi was its first recipient, and he was given $25,000. When Fermi died, the award was named in his honor and the amount fixed at $50,000. The choice of the recipient was left largely to the advisory committee of the Atomic Energy Commission. By 1964, seven of the nine members of this committee had received the Fermi award. In 1956, it was given to John von Neumann; in 1957, to Ernest O. Lawrence; in 1958, to Eugene Wigner. Wigner had resigned from the advisory committee late in 1956 and was reappointed in 1959. In 1959, Glenn Seaborg was honored. In 1960, a deadlock developed over whether the award was to go to Hans A. Bethe or Edward Teller. No one got the money in 1960, but in 1961 Bethe was the recipient and Teller made the list in 1962. In 1963, J. Robert Oppenheimer received the award at a White House ceremony. Some members of Congress were infuriated at the award because Oppenheimer had earlier been accused of Communist affiliations. They took the opportunity to point out that a $50,000 tax-free award was a great deal of money and that the advisory committee had been awarding itself. In 1964, the Fermi award was cut to $25,000 and given to Vice Admiral H. G. Rickover.

The Fermi award sequence may have indicated a conflict of interest to some nonscientists. But it was not the kind of flagrant case that enmeshed Dr. Henry Welch, one-time head of the antibiotic division of the United States Food and Drug Administration. While he was a salaried employee of the government with a responsibility for protecting the public against abuses in drug marketing, he became editor of a journal, *Antibiotics and Chemotherapy*, that argued for drug prescription practices that many scientists opposed. Welch collected more than a quarter of a million dollars as an editor before he was forced to resign from government service during the summer of 1960.

During the American Civil War there were scandals in the military procurement program, and the conflict-of-interest laws were drawn up as a result. There were no serious challenges to the wisdom of these laws until after World War II. In 1960, the New York Bar Association published an exhaustive study of the laws and found that scientists had the greatest difficulty in conforming to them. At times, scientists were acting as consultants, sifting research proposals, or assessing research results for private industry while simultaneously doing research under contract to a government agency. Private gain or collusion was not involved. There existed only one set of people who could both advise and also serve the scientific needs of the government. During the early 1960's, the first ten universities in amount of federal money received for scientific research accounted for 38 percent of all such federal support; the same universities also gave 37 percent of all advisers used by the government to review research proposals. The Bar Association recommended a rewriting of the conflict-of-interest laws with enough leeway to allow exceptions in the national interest.

The national interest is not always clear to scientists. In 1969, experts in geology, geophysics, and petroleum engineering turned down requests to testify for the State of California in its half-billion-dollar damage suit against four oil companies allegedly responsible for the oil spillage off Santa Barbara. The men, who were university professors, also had close ties with the oil industry and did not wish to risk losing grants and consulting arrangements. In the same year a report of a House of Representatives Committee on Government Operations detailed the Department of Agriculture's deficiencies in administering the Federal Insecticide, Fungicide, and Rodenticide Act (FIFRA)." They claimed "serious conflict-of-interest questions in the Agricultural Research Service's appointment of consultants who either worked for or were consulting for Shell Chemical."

Scientific organizations have been silent about the conflict of interest. In December 1964, such nonscientific organizations as the American Council on Education and the Council of the American Association of University Professors joined in issuing a statement about the prevention of conflict of interest in government-sponsored research at universities.

Though scientists may be excused from conflict-of-interest laws when national security is involved, they are not exempt from other monetary responsibilities. The Washington correspondent of the *New York Times,* John W. Finney, reported "one brokerage house in Washington has unusually active accounts with some scientists employed by nonprofit research groups serving the military departments." In 1963, the National Academy of Sciences report *Federal Support of Basic Research in Institutions of Higher Learning* claimed that the investigator "assumes a major responsibility in accepting federal funds and has an obligation to account for their proper use." In the October 14, 1966, issue of *Science,* the publisher, Dael Wolfle, wrote in the editorial "Academic Responsibility": "University presidents have generally understood the importance of keeping control at the institutional level. But scientists often have not, and some have failed to recognize the need that there be public confidence that public funds are used prudently and honorably. They have talked much of academic freedom without accepting the correlating requirement of academic responsibility."

The first clear-cut case (rather than rumor) of questionable use of government funds by scientists came to light in the early 1960's and was handled with finesse and dignity by the scientists who felt implicated. Auditors of the National Science Foundation found discrepancies in the use of government funds awarded by them to the American Insitute of Biological Sciences. The latter was organized in 1947 as a part of the division of biology of the National Research Council in order to upgrade biology. In 1954, the AIBS was established on an independent basis with fifty affiliate societies having a total of about 80,000 members. Money to support the activities of the AIBS came from an assessment of one dollar per member of the affiliate societies plus a basic fee for each society. Hiden Cox, formerly a professor at Virginia Polytechnic Institute, who was deputy executive director, became director. Public money began to flow to the organization, from $56,000 in 1956 to $3,000,000 in 1962; the headquarters staff in Washington rose from seven to seventy. The National Science Foundation provided most of the new funds for the growing AIBS, but the Foundation did not audit the AIBS books until the fall of 1962; the Foundation always sought to minimize its intrusion into the affairs of its recipients.

Perhaps AIBS entered its trouble phase when its controller died and there was difficulty in finding a suitable replacement. At almost the same time a new comptroller was installed at the National Science Foundation.

The American Institute of Biological Sciences was charged in 1963 with using large amounts of NSF funds for purposes not intended. The NSF charged the AIBS with taking excessive indirect costs for administering NSF grants, listing such nonacceptable items as a few thousand dollars for entertainment and a lesser sum for international travel, collecting interest on invested NSF funds and not refunding the interest to the government, and failing to obtain NSF approval for the use of funds realized from the sale of publications. Under NSF regulations, the latter funds are held in escrow and used only for purposes approved by the NSF; AIBS had placed this money into a general fund. NSF director Alan T. Waterman emphasized there was "no evidence of personal gain" resulting from the financial operations.

James D. Ebert, director of the embryology department of the Carnegie Institution of Washington became president of the AIBS just after the financial irregularities had been discovered by an NSF auditor. Ebert stated in a letter to AIBS board members that there was no question that the organization was morally wrong in using funds awarded for one purpose for some other purpose. The board of directors of the AIBS sent requests to individual members of the affiliated societies asking for at least $10 per year, beginning 1963, for repayment of the debt to the government and the maintenance of basic operations. The first letter, sent January 1963, drew four thousand replies with pledges and cash amounting to about $34,000. By the end of February, there were twelve thousand replies with contributions, as well as membership fees, totaling $110,000. Negotiations between NSF and AIBS continued; the funds required were considerably reduced on the basis of more careful auditing. In time, the former executive director secured an academic post at Long Beach State College in California. Apparently the only financially hurt individuals were thirty-six members of the AIBS staff who were laid off without severance pay and with only two weeks' notice.

In Italy, the money mistakes of scientists have not always had as happy an ending. During the summer of 1963, Felice Ippolito, a professor of geology and the head of the Italian Atomic Energy Agency, was arrested for misappropriating funds, as well as for giving contracts to a company with which his father was connected. In October 1964, he was convicted of embezzling more than $1 million and was sentenced to eleven years of imprisonment. Six other defendants, including his father, were given prison sentences. During the trial it was revealed that he had cashed social security benefits under false pretenses and that the Atomic Energy Agency had paid his bills for private trips. On February 4, 1966, a court of appeals in Rome reduced the eleven-year sentence to five years and three months and completely absolved the six other defendants in the case involving Ippolito's father. Critics of the trial and indictment continue to insist that atomic energy programs cannot be successfully administered within the legal requirements of the horse and buggy days.

During the spring of 1964, Giordano Giacomello and Domenico Marotta, the current and former heads of the Italian National Public Health Institute respectively, were arrested along with some other administrators and charged with misappropriating funds. Italian scientists rose in almost unanimous protest. More than four hundred research scientists at the National Public Health Institute sent the Italian president a document of support for the 78-year-old Marotta. More than seventy-two professors, including three Nobel prize winners, sent a letter to the Italian prime minister, praising the "spirit of sacrifice" of Giacomello and Marotta. Almost all blamed the archaic laws of Italy; for example, there was no provision in the law for having laboratory coats laundered at state expense and so using funds for such a purpose broke the law. The Minister of Health drafted a law to give the National Public Health Institute a greater financial freedom. On July 25, 1965, Marotta was found guilty and sentenced to six years and eight months in prison. Under Italian law he cannot be imprisoned after his eightieth birthday, and this will occur by the time his appeal is handled.

Accusations of financial irregularities by scientists are destined to become more frequent. The money spent on science and by scientists is becoming more noticeable and will stimulate the airing of real and imagined grievances. Newspapers seeking interesting stories will undoubtedly seek out the possibilities. During the summer of 1964, the Waldemar Medical Research Foundation, a small institution on Long Island, New York, was accused by the Long Island daily newspaper *Newsday* of misuse and mismanagement of publicly donated funds. The newspaper published their charges in many articles over a couple of months. The Nassau County district attorney began to investigate. In Japan in 1967, the newspaper *Asahi Shimbun* "exposed" the U.S. Army, which had been making research funds available to Japanese scientists from 1959 on as well as giving support to the Japan Physics Society for its International Conference on Semiconductor Physics. Earlier the paper had been instrumental in causing the resignation of Professor Hideo Itokawa, who had headed his university's Institute of Space and Aeronautical Sciences; in a series of articles, the newspaper had also questioned the management of Japan's rocket program. In 1970, the *Chicago Tribune* revealed that federal auditors had found discrepancies in the allotment of money received for overhead expenses by Hektoen Institute, a private agency administering research grants for the public Cook County Hospital.

Even if pure science became an international fiscal responsibility with research projects being funded over longer periods of time, money mistakes would not necessarily be minimized. Fiscal irresponsibility could still be possible even if money for pure research came from all countries, based on some kind of formula involving population and gross national product. To minimize monetary blunders, there must be a deeper sense of commitment to pure science and its goals, as well as capable administration of scientific research.

Applied science and development, as well as borderline cases between pure and applied science, will inevitably be more susceptible to money mistakes and controversies. The work of Robert J. Code, a professor of renal medicine at the University of Florida's Medical School, illustrates the point. He developed Gatorade, named for the university's football team, the "Gators," in 1965 after a medical school security officer, who was also a freshman football coach, asked him why players lost as much as fifteen pounds during a game. Code spent a very small amount of a grant he had received from the National Institutes of Health, as well as his own money, to study the problem, and as a result he produced Gatorade, a solution of chemicals and flavored water. The Stokely-Van Camp Company purchased patent rights and then Code's troubles began. He was deluged with threats—from the federal government for alleged violation of the terms of a grant; from the state university board of regents for alleged breach of his contract, which stipulated that all inventions developed by employees of the State of Florida belong to the state, unless they are waived; and from other inventors for alleged patent piracy.

Scientists and the public will need to have greater understanding if research is to continue to have public support and confidence. Both the public and the scientists will need greater financial sophistication—the public so that they can appreciate the role of public money in research ventures, and the scientists so that they can be financially responsible as researchers and administrators. It may be that scientific organizations, along with other large organizations, will need to employ public relations experts to insure that the mass media will continue to look with favor at scientific activities.

SELECTED REFERENCES

Goran, M. "Scientists Really Like Money," *Engineering Opportunities,* 2 (November, 1964), 14, 28; reprinted in *The Chemist* (February, 1966).

McElheny, V. K. "Research Climate in Italy," *Science,* 145 (August 14, 1964), 690–93.

———. "E. B. Chain Accused of Contempt of Italian Judiciary," *Science,* 150 (December 17, 1965), 1573–75.

Scientific American, 213 (September, 1965), 82.

3
FIRST LOYALTY

Scientists in the United States have for the most part had friendly relations with politicians. Thomas Jefferson and Benjamin Franklin began this rapport by pioneering in both government and science. After World War I, and particularly after World War II, science became a necessity for the industrial state. Now, through government, science can try to solve its problems. Government controls will undoubtedly be involved in science's future communication, research priorities, and education.

Soon after World War II, scientists in the United States learned to be a persuasive force in politics. As amateurs in politics, they lobbied and testified before Congressional committees in support of civilian control of atomic energy. Later they became more adept. They were very active in the 1964 election campaign. "Scientists and Engineers for Johnson" included ten Nobel-prize winners as well as the former science advisers to Presidents Eisenhower and Kennedy. Thirty-three Nobel laureates endorsed Johnson at a huge rally held in Washington, D.C. When President Johnson saw a sign in Albuquerque, New Mexico, with the inscription "New Mexico Scientists and Engineers Welcome LBJ," he responded that scientists and engineers were "about the best support I have." When they became disenchanted with Johnson over the Vietnam war, most of the scientists concerned used the public forum singly or in groups to indicate their stand. Late in 1967, some of them formed Scientists and Engineers for McCarthy to bolster the campaign of Senator Eugene J. McCarthy. Republicans among the scientists and engineers have also indicated their political preferences by forming similar committees.

Antiscience legislation in the United States has never had much support. Such an aberration as Tennessee's law against the teaching of organic evolution was viewed tolerantly and humorously. At least it provided the Dayton, Tennessee trial (and circus for some) in 1925 with a conflict between attorneys William Jennings Bryan and Clarence Darrow. Only very minor attempts were made by scientists to remove such legislation. The University of Tennessee and other major institutions within the state ignored the statute. However, in 1967, the Tennessee ban was repealed, and in 1968 the United States Supreme Court ruled unanimously that state laws prohibiting the teaching of evolution in public schools are unconstitutional. Early in January 1970, a bill to repeal Mississippi's forty-four-year-old law against the teaching of evolution was defeated in the state's house of representatives by a vote of 70 to 42. Mississippi was the last state which still had a law against teaching in public schools that man "ascended or descended from the lower order of animals." In December, 1970 an appeal court overturned the anti-evolution law.

In other countries scientists have sometimes not fared as well in their relations with government. At least twice within the twentieth century, scientists were not able to counteract governmental policies at odds with the results of modern science; twice they were inept and powerless when a state promulgated doctrines contrary to scientific truth. In one instance the impotence of the scientists was surprising because the state, Germany, had nourished and cultivated science; in the other, the U.S.S.R., the inability of scientists to mount an attack was not unexpected because of the weak scientific tradition in the country.

The science of biology had been sustained in Russia largely through the work of the talented physiologist Pavlov. After the revolution, there came to be a school of young Russians interested in the science of heredity. Russian scientists began to swarm to the specialty because it held much promise for understanding the nature and history of organisms, and for the improvement of agriculture, animal husbandry, and even human beings. Although the communist godfathers Marx, Engels, and Lenin had commented on science, they had formulated no specific doctrine for the science of genetics. Marx and Engels had lived before the founding of modern genetics, while Lenin was more the political and social critic. The communist doctrine did extol the virtues of the common man and the possibility of his salvation as a result of a suitable environment; it was more inclined to dismiss heredity as an unimportant agent in causing human behavior. In their perspective the millennium could be reached by all in the perfect society.

At the beginning of the nineteenth century, the French scientist Lamarck had developed the thesis of the inheritance of acquired characteristics. Organisms, the thesis went, developed new attributes as a result of the environment, and these could be passed on from generation to generation. Other scientists denied this view. At the beginning of the twentieth century, both schools were ready to adopt the invisible subcellular unit the gene as the agent transmitting characteristics. The gene could be altered by such natural environmental influences as heat and x-rays in the process called mutation, and these changes were generally deleterious. To accent environment as opposed to heredity, the Soviet theorists were inclined to adopt the theme of the inheritance of acquired characteristics. They needed but a scientific herald. He appeared in the person of Trofim Denisovich Lysenko.

Lysenko was born in the Ukraine in 1898. He was graduated from the Kiev Institute of Agriculture and almost immediately thereafter began a series of experiments in plant breeding. These, he later said, convinced him that such "acquired" characteristics as resistance to weather changes are inherited. This conclusion put him in direct opposition to biologists everywhere who believed that acquired characteristics are not inherited. Lysenko managed to influence dictator Joseph Stalin more than did the classical geneticists. Slowly the classical geneticists began to disappear from the Russian scene. In about 1933, the geneticists Chetverikoff, Ferry, and Ephroimsen were sent separately to Siberia; Levitsky was sent to a camp in the European Arctic. The erosion was slow; the opposition to Lysenko evaporated away. In 1936, Agel disappeared. In the same year, the

Medicogenetical Institute, with a staff of two hundred scientists, was dissolved; one of the charges was that they attempted to exalt heredity in the environment-versus-heredity controversy. The founder and director of the Institute, Solomon Levit "confessed" scientific guilt; he later admitted to American Nobel-prize winner H. J. Muller that the confession had been false and that it had been demanded as a sign of loyalty to the Communist Party. Nonetheless, Levit was not heard from after these incidents.

A public debate between the supporters of Lysenko and the classical geneticists was arranged by the Communist Party in December 1936. Although the classical scientific workers were the apparent victors, both the Soviet press and the chairman of the meeting adversely criticized them. The Seventh International Congress of Genetics, scheduled for the summer of 1937 in Moscow, was abruptly called off. In 1939, Edinburgh, Scotland, was host to this convention, and all forty Soviet geneticists were forbidden to attend. In the same year, Nicolai Ivanovich Vavilov, who had been forced to send a letter of resignation to the Edinburgh Congress, lost his posts as president of the Lenin Academy of Agricultural Sciences, head of the Institute of Plant Production, and head of the Institute of Genetics. All the posts were taken over by Lysenko. Within a year Vavilov was arrested, charged as a British spy, and sent to Siberia, where he died in 1942. His brother, physicist Sergei Vavilov, head of the U.S.S.R. Academy of Science, followed the party line in 1948 when he removed Soviet physiologist Orbeli, morphogenesis expert Schmalhausen, and geneticist Debinin from the Academy.

After World War II the Soviet Union, a member of the victorious Allies, may have felt the need for another showing of Lysenko versus the classical geneticists. (In 1939, the Soviets had staged another debate between the supporters of Lysenko and the classical geneticists, but the latter had already been weakened by deportations and executions.) In August 1948, the Lenin Academy of Agricultural Sciences held such a session. The Lysenko adherents repeated their claims and emphasized other new ones. They held that statistics had no place in biology and therefore they could ignore the work of Gregor Mendel, the founder of classical genetics. When asked why experiments performed in and out of the Soviet Union could not duplicate the work of Lysenko, the reply was that the proper technique had not been employed. According to the record of proceedings of this meeting, one attendant had the courage to say: "I want to make a personal request of Trofim Denisovich. Trofim Denisovich, instruct your organization to issue a comprehensive manual on how to train plants, on how to alter them. Teach us; we too want to learn and if your methods prove effective, we will accept them." An instruction manual was never issued.

Toward the close of the meeting, Lysenko was asked about the attitude of the Communist Party. He replied that the party had already approved his ideas. Evidently this was the signal for all to fall in line; Lysenko's opponents recanted their prior arguments and promised to support the party and government doctrines. The proceedings of this meeting report a member of the Academy Zhukovsky saying: "The speech I made the day before yesterday, at a time when the Central Committee of the Party had drawn a dividing line between the two trends in biological science, was unworthy of a member of the Communist Party and of a Soviet scientist." S. I. Alikhanian said: "From tomorrow on I shall not only myself, in all my scientific activity, try to emancipate myself from the old revolutionary Weismann-Morganian views, but shall try to reform all my pupils and comrades."

It has been argued that the spectacle of the Soviet geneticists capitulating to the regime was comparable to that of Galileo before the Inquisition. But there are vital differences. During Galileo's time, modern science was an infant that was defended and practiced by a few; in the early twentieth century, modern science was a much stronger youngster, with a background of growth and tradition. In Galileo's time, the Church permeated Europe and almost all European intellectuals were within its framework; at the time of the Soviet purge, Communism had been in power a relatively short time and many thinkers were not under the influence of its ideology. Galileo was an established, elderly gentleman when he recanted; many of the Soviet scientists were young and not fully arrived in accomplishments. Galileo could believe that the Copernican system had no physical reality and was a convenient face-saving myth; the

Soviets had to relinquish well-established ideas in favor of those already proved wrong.

Soviet scientists did not even begin to resist the supremacy of political power. They allowed it to form scientific truth. The record shows extermination of possible opponents by the political power and the meek acquiescence by those remaining. Even during Krushchev's 1954–64 reign, Lysenko was the director of the Institute of Genetics and editor of *Acrobiology*. There may have been some degree of freedom allowed in other areas of Soviet life during this time, but it was minimal.

Only after Krushchev's downfall was Lysenko ousted. Early in 1965 he was removed as the director of the Institute of Genetics of the U.S.S.R. Academy of Sciences. M. V. Keldysh, the president of the Academy of Sciences said at the time: "The exclusive position held by Academician Lysenko must be submitted to free discussion and normal verification. If we create in biology the same normal scientific atmosphere that exists in other fields, we will exclude any possibility of repeating the bad situation we witnessed in the past." Early in 1966, the U.S.S.R. issued a report by a commission of scientists charging that the claims of Lysenko had been frauds based upon falsification.

The Soviets had been so intent upon imposing the views of Lysenko that they had missed the chance to challenge the Nazis when the Germans had imposed their erroneous doctrine of race upon the flourishing science of Germany. In 1936, when the first debates on genetics were permitted, the Nazis had been in power a few years and had not yet made a pact with the Soviets.

The false notion of race supremacy, held even today by many in spite of the achievements of all races in all fields of knowledge, gained some kind of respectability as a result of the publications of French Count Joseph Arthur de Gobineau and Englishman Houston Stewart Chamberlain. Both of these racists claimed racial purity as a foundation for national superiority. The Nazis adopted the views of these foreigners.

The established scientists of Germany thought that Hitlerism was equivalent to a shower that would soon be over. One said that a few Polish Jews would be relieved of their posts and the race business would be done. When events proved these beliefs to be entirely erroneous, the scientific community in Germany was impotent.

Not long after Hitler had assumed power, the Nazis removed one of the most important figures in German science. Fritz Haber had been a veritable hero during World War I. His successful nitrogen-fixation process had given staying power to Germany. As head of the German Chemical Warfare Service, he had instituted poison gas attacks. After the war he had valiantly attempted to remove gold from the oceans in order to pay the reparations demanded by the Allies. This Nobel-prize winner was the head of the Kaiser Wilhelm Institute for Physical Chemistry and Electrochemistry, one of the many research institutes that were bringing acclaim to Germany as a country fostering science.

On Friday, April 21, 1933, Haber called his chief assistants, both of Jewish ancestry, to inform them that he had received a message from the Ministry of Education to the effect that work at the Institute could not continue unless the many Jews employed there were removed. (During the years before the Nazi accession to power, many of the Jewish research workers had been unable to secure more lucrative posts at a university or industrial laboratory and had remained at the Kaiser Wilhelm Institute.) Haber had been told to come to the Minister's office to work out a reorganization. "If you go," said Herbert Freundlich, one of the two chief assistants, "take my resignation with you." Haber advised a rewriting of the resignation letter because according to him nothing could be gained with strong language.

At the Education Minister's office Haber was again told that the new regime could not have the intolerable situation of employing a tremendous number of Jews at his Institute. Haber, a converted Jew, requested his own dismissal. The official backed away, saying Haber was a famous man and was not involved in the order. They flattered him with the truth, and he proceeded to deal with them.

On Monday, April 24, the acting director of the Kaiser Wilhelm Institutes returned and after investigation reported that all that had transpired was a mistake. Only the lesser assistants were meant; he brought back the resignation letters of the chief assistants.

A few days later, Bernhard Rust, the Minister of Education, spoke at a public meeting. He said that he had received a letter from a German Jew, a chemist, who had written to say he had always chosen collaborators on the basis of their scientific qualifications, not their ancestry. This, said Rust, was an intolerable situation.

After his letter had been published in the newspapers, Haber went to Rust's office, where he was received by an assistant. There had been a possibility of an understanding, Haber was told, but after his letter there was no possibility of "collaboration." Haber was also warned to behave like the successor to Bismarck; the latter, Caprivi, had not brought any information to the newspapers when he was dismissed.

Some forward-looking German scientists sought to head off the debacle. Carl Bosch, the scientist who applied the Haber process of nitrogen fixation, came to Berlin intending to organize a strong resistance of non-Jewish German professors. He wanted Max Planck to be the leader, but Planck was in Italy. He approached physicist Max von Laue who suggested that a younger man be chosen; besides, von Laue's health was failing and he was not a public speaker—he stammered. Physicist Erwin Schroedinger also gave excuses.

Max Planck did get to speak to Hitler about the situation, and Hitler became red in the face as he said, "If you still want respect in Germany, stop speaking like that." A student of noble heritage in Haber's laboratory did attempt to organize resistance among his friends, but he returned about ten days after he had started with the comment, "We cannot draw our swords for the Jews." Psychologist Wolfgang Koehler came to his classes saying "Heil Hitler" and added, "I only do this because I am ordered by the government." When he was told to stop these latter remarks, he failed to come to classes.

The great bulk of non-Jewish German scientists were raised to respect truth, but they adjusted quickly to the government. On February 2, 1933, a few days after Hitler had assumed power, anthropologist Eugene Fischer gave a public lecture at

Kaiser Wilhelm Institute. He stated that the most celebrated Germans had come from regions in which the least pure German lived, such as Saxony and the Rhineland. He openly stated that gifted men arise from a mixture of races, such as Jews with Aryans. Within a few weeks, this statement was forgotten as Fischer came to accept, at least on the surface, the race doctrine of the Nazis.

The resistance never came to fruition, although a few seeds of revolt were present. On the first anniversary of the death of Fritz Haber—he had died in exile in 1934—the Kaiser Wilhelm Institute for Physical Chemistry, the German Chemical Society, and the German Society of Physics made plans to honor him. The Minister of Education forbade all officials and teachers to attend the memorial services. But on January 29, 1935, about five hundred scientists gathered at Dahlem, a suburb of Berlin and the site of the Institute. Max Planck opened the eulogy with the Nazi salute and began: "We reward loyalty with loyalty and pay our earnest tribute at this moment to the German scholar and German soldier Fritz Haber." About a year later, in response to an attack upon Einstein and other Jewish scientists in the scurrilous *Voelkischer Beobachter,* organ of the Nazis, six Nobel-prize winners came to the defense of theoretical physics. But it was evident by that time that German scientists identified themselves with Germany, not with science. The majority of them viewed the Nazi episode as directed against Jewish scientists rather than science.

The same attitude that an outburst against scientists and scholars was not directed against science was prevalent among world scientists in 1966. During the summer of 1966, the Ongania dictatorship in Argentina raided the national universities and in some cases even did physical harm to scientists and scholars. Professor Carlos Varsovsky, director of the radio observatory in La Plata, received serious head wounds; Felix Gonzales Bonorino, the best-known geologist in the country, was hurt. At the University of Buenos Aires, nearly 200 of the 350 members of the science faculty resigned. Rolando Garcia, dean of the Faculty of the Exact and Natural Sciences, joined the International Union of Geodesy and Geophysics in Geneva, Switzerland. Juan Roederer, a cosmic-ray researcher, accepted a post at the University of Denver. By early 1967, the physics and inorganic chemistry departments of the school were not operating; the Applied Mathematics Center had closed.

Present aberrations of governments are not too well publicized. A letter to *Science* revealed that an Israeli entomologist planned to attend the thirteenth International Congress of Entomology in Moscow, August 2–9, 1968, and his arrangements had been confirmed by the secretariat of the Congress. Yet he received neither a Soviet visa nor an explanation; his telegrams and letters to the Congress and the Soviet authorities were unanswered even though the answers were in some cases prepaid. The manuscript of his paper, earlier accepted, was not allowed to be read. In June 1968, an American biologist of Jewish origin used the same letter to *Science* technique to report his failure to receive a promised visa to enter Poland, and to document the persecution of Polish intellectuals and scientists of Jewish origin. The rumor circulates that the Soviet authorities punished the noted physicist Andrei D. Sakharov in some way after the circulation of his essay "Progress, Coexistence, and Intellectual Freedom," published in the United States in the *New York Times,* July 22, 1968, and later published in book form. Many scientists in Czechoslovakia fled after the Soviet invasion in 1968. Nieh Jung-chen, considered the father of mainland China's atom bomb, has been denounced in wall posters there. A 1968 editorial in the Shanghai Communist Party newspaper *Wen Hui Pao* criticized the scientific community for hiding "away in the typhoon shelter against ideological struggle" during the Chinese cultural revolution.

Fiscal support of pure science as an international venture would remove some of the still prevalent national tendencies to use science as a political pawn or to hinder science and its development. Those countries rich in wealth, population, or land areas could not stifle a theory of organic evolution or the concept of relativity. The Dutch Reformed Church in South Africa might still ban the teaching of evolution, but most people would rely on the critical faculties of world scientists to render judgment on the worth of a scientific idea.

SELECTED REFERENCES

Caspari, E. W., and Marshak, R. E. "The Rise and Fall of Lysenko," *Science*, 149 (July 16, 1965), 275–78.

Goran, M. "Swastika Science," *Nation*, 148 (June, 1939), 641.

———. "The Case for Politics for Scientists," *Industrial Research*, 6 (March, 1964), 27–28.

———. *The Story of Fritz Haber*. Norman: University of Oklahoma Press, 1967.

"Governments as Patrons of Science," *Nature*, 224 (October 4, 1969), 2–3.

Greenberg, D. "Venture into Politics: Scientists and Engineers in the Election Campaign," *Science*, 146 (December 11 and 18, 1964), 1440–44, 1561–63.

Joravsky, D. *The Lysenko Affair*. Cambridge: Harvard University Press, 1970.

Medvedev, Z. A. *The Rise and Fall of T. D. Lysenko*. New York: Columbia University Press, 1969.

Muller, H. J. "Science in Bondage," *Science* 113 (January 12, 1951), 25ff.

4

SPIES

Not only have scientists capitulated to governments and accepted conclusions contrary to objective research, as in Germany and the Soviet Union, but they have also been known to serve as spies. Such activity may give some immediate material advantage to the spy, but in the long run it damages the profession's image and hardly contributes to the solution of the problems of science which need governmental assistance. Indeed, it may adversely affect the allocation of money for research.

An American, Benjamin Thompson, later Count Rumford, is the most famous spy, and perhaps the only one, of pre-twentieth-century science; he was also a very accomplished scientist. Thompson's experiments led to the downfall of the caloric theory, the idea that heat was an intangible fluid present in definite quantities in materials; caloric was said to be imponderable, self-repellant, and weightless. Thompson was one of several who showed that continual mechanical action produced continued heat—there evidently seemed to be no limit to the amount of caloric. He was the first to find what are today called convection currents, moving masses of material carrying heat. Thompson's practical work included a study and subsequent improvement of textiles, nutrition, gunpowder, stoves, fireplaces, and lamps. He was a promoter of scientific research— he founded the famed Rumford medals given in England and the United States. He encouraged the spread of scientific knowledge by creating the Royal Institution of London and by employing Humphry Davy to direct it. As a city planner he installed the English Gardens of Munich, a park in the center of a metropolitan community. As a social reformer he established workhouses for the beggars of Munich. In short, Benjamin Thompson was a man of many accomplishments.

He was born March 26, 1753, on a farm near Woburn, Massachusetts. His father died when he was quite young, and his mother remarried. Thompson went to work when he was thirteen years old and he held apprentice positions in dry goods shops for about four years; then the physician at Woburn accepted him as an apprentice. This too was terminated and Thompson became a school teacher in Concord, New Hampshire. Here, at the age of nineteen, he married a thirty-year-old rich widow, less than four months after he had met her. Success followed quickly: Benjamin Thompson received a major's commission in the New Hampshire militia. His commission was a guise to cover his real work as a spy for the Royal Governor of the State; the latter wanted information about dissenters as well as about the nature of the rebellious sentiment in the area. The people of Concord, suspicious and resentful of Thompson's openly stated royalist sentiments, summoned him to a Committee of Safety on the charge of being unfriendly to the cause of freedom. The case was dismissed, but the more zealous partisans formed a group to tar and feather Thompson and ride him out of town. He escaped before the mob action could materialize, which meant he had to leave his wife and child; he was never to see his wife again.

Thompson went to Boston and then back to Woburn. During this time he had the gall to offer his services to Colonel George Washington. He continued on as a royalist sympathizer and as a royalist spy. On May 6, 1775, he sent a letter written in invisible ink to the British which the Americans never detected; he was knowledgeable enough to use gallo-tannic acid made by soaking powdered nutgalls in water. Because of his known sympathies, he was again called before a Committee of Safety, this time in Massachusetts. Again nothing was proven, but not long afterward Thompson arrived in London.

He must have been impressive, enterprising, and persuasive. He became the private secretary to the Secretary of State for the Colonies, Lord George Germain, and retained this position for five years. As part of his duties, he accompanied Admiral Hardy's fleet in its maneuvers during the summer of 1779. He was there ostensibly to arrange gunnery experiments and make other scientific studies, but he sent exhaustive reports on other matters to Lord Germain. He was an accomplished spy not only for the British, but also for the French, to whom he evidently sold his talents. The French spy LaMotte, caught by the confession of a confederate, never identified his British contact, save for the phrase "friend in a certain office." This "friend" was said to be Thompson. Soon after LaMotte's trial, Thompson left his position as Undersecretary of State for the Northern Department with the responsibility for recruiting, equipping, and transporting the army and navy. He returned to America.

At the end of the American Revolution, Thompson turned his talents as a spy, patronizer, administrator, and entrepreneur toward the continent of Europe. He arranged with Sir Robert Keith, the British ambassador in Vienna, to be an informer for him. At the same time he so charmed the monarch of Bavaria, Elector Karl Theodor, that he was appointed a military aide and adviser to the Bavarian. As he prospered in his post at the Bavarian court, he diminished his activity for the British. They evidently thought he was withholding information because when he returned to England for a visit after eleven years in Bavaria, he was attacked by "highwaymen" and a trunk of papers was taken from him. (Thompson did not hesitate to use a similar technique to gain his own ends. According to the director of the Poor People's Institute in Munich Mr. Piaggine, Thompson hired men to beat the director into submission when persuasion had failed to win him over to Thompson's plan for a workhouse.)

Benjamin Thompson, the spy, worked all sides of the street for the benefit of Benjamin Thompson. This procedure was beneficial in at least one specific instance in addition to the long-term one of sustaining Benjamin Thompson to do scientific work. In 1796, he gave both to the Royal Society of London and the American Academy of Arts and Sciences in Boston $5,000 with which to establish prizes for the outstanding scientific research in the fields of heat and light. The American Academy did not get to award a prize until forty years later; the English society gave its first Rumford prize to Benjamin Thompson, Count Rumford in 1802.

The scientist spies of this generation have not the enormous number of achievements of Rumford. In comparison, the spying achievements of Bruno Pontecorvo, Klaus Fuchs, and Allen May Nunn are minimal in number. However, in terms of the impact of their spying, the present generation outstrips Rumford by far.

Allen May Nunn was a British physicist who admittedly gave information about the atomic bomb to the Soviets. In his court defense he claimed that he was not a spy but a humanitarian. He said he had divulged no secrets but rather had helped out an ally. The Association of Scientific Workers of Great Britain did not seek to justify his breach of the Official Secrets Act, but they deplored the ten-year sentence he received as excessive. Nunn served six and three-quarters years of his term, and when he was released on December 29, 1952, he issued a handwritten statement to the press which said, "I myself think that I acted rightly and I believe many others think so too." He said that he had been wholeheartedly concerned with victory over Germany and Japan. Within six months after his release, he married Vienna-born Dr. Hildegarde Broda, an assistant county medical officer in Cambridge, and resumed scientific work.

The British and American citizens with scientific training who spied for the U.S.S.R. before and during World War II were largely applied scientists and engineers. Chemist Harry Gold was working in a Philadelphia hospital when he was arrested. He had stolen industrial as well as atomic data; one Eastman Kodak chemist had given him color-film information for relay to the Soviet Union. Julius Rosenberg, executed by the U.S. for spying, along with his wife, had taken a degree in electrical engineering at the City College of New York. French-Canadian Raymond Boyer was a chemist who gave Soviet agents the formula for a new method of producing a valuable explosive.

The case in 1950 of Dr. Emil Julius Klaus Fuchs was more significant. The thirty-eight-year-old German-born physicist held important positions in the British atomic establishment, with easy access to American installations. He was arrested for giving American atomic secrets to the U.S.S.R. in 1945 and for giving British secrets to that country in 1947. As in the Nunn situation, ideology rather than money was the motivation; Fuchs received but $280 from the Russians. In February 1950, Fuchs was stripped of his British citizenship and was given a sentence of fourteen years for "grossest treachery." When released in July 1959, after nine and one-half years in prison, he went to communist East Germany.

In 1950, Bruno Pontecorvo disappeared after the disclosure of the atomic spy ring of Fuchs and company. Pontecorvo, a first-class scientist, vanished with his Swedish wife and their three sons while supposedly on a vacation in Italy. The United States immediately impounded the $18,000 he had been awarded for his share of patent rights on a process for slowing neutrons in an atomic reactor. Early in 1955, Pontecorvo held a news conference in Moscow to announce that he was then a citizen of the U.S.S.R. and that he had won a Stalin prize in 1952.

In recent years some American anthropologists have served as spies for the Central Intelligence Agency. In 1967, anthropologist Ralph L. Beals of the University of California at Los Angeles, a former president of the American Anthropological Association, reported that it could be stated "with considerable confidence" that agents of the intelligence branches of the U.S. government, particularly the CIA, had posed as anthropologists and that some trained anthropologists, representing themselves as anthropological researchers, have been affiliated with the CIA. He reported that one U.S. anthropologist is believed to be a full-time agent of a foreign power.

Because of these revelations, the opportunities for American scientists to do research in other countries have been cut down considerably. American researchers are regarded with increasing suspicion in Latin American countries. During the summer of 1968, Brazil imposed frustrating restrictions on oceanographic research by non-Brazilians in Brazilian waters. The University of Miami's Institute of Marine Science failed to get clearance for a project; Louisiana State University could not research sandbar formation in the mouth of the Amazon. In 1967, Brazil withdrew from an arrangement which was going into its third year—the Cornell-Brazil project—and the Brazilians wrote to Cornell University: "How can one maintain and justify a relationship with an institution—the university in the United States—which permits itself to be transformed into the instrument of a security agency which today is internationally known as the instigator of dictatorial coups?"

When members of the scientific community give their first loyalty to science, the incidence of spying will decline. The individual researcher or teacher, subject to pressures from nation, church, and home, must be able to withstand them and yield only to professional responsibilities in harmony with scientific humanism. The chances for him to carry out his responsibilities without outside pressures would be greater if funds for pure science were collected internationally and distributed in like manner.

Pure scientists have always been proud of their international brotherhood, and until recently the cost of maintaining this point of view has not been high. Now, however, international conferences and cooperative ventures must generally be supported by contributions from one or more of the powerful nations. The United Nations Educational Scientific and Cultural Organization depends upon large amounts of money from the United States, and nationalistic pressures from several countries are felt within the organization. The continued international union of scientists for the support and development of pure science would do better coming from a movement by scientists and their societies.

SELECTED REFERENCES

Barr, E. S. "Count Rumford," *American Journal of Physics,* 32 (April, 1964), 292–96.

Brown, S. C. *Count Rumford, Physicist Extraordinary.* New York: Doubleday, 1962.

Sparrow, W. J. *Count Rumford of Woburn, Mass.* New York: Crowell, 1966.

West, R. *The New Meaning of Treason.* New York: Viking, 1964.

Wilson, M. "Count Rumford," *Scientific American,* 203 (October, 1960), 158–68.

5
SOCIAL RESPONSIBILITY

If scientists wish to have more and better help from nonscientists in solving the current and future problems of the profession, scientists must dispel the current false notion about them. The American public seems to have accepted an erroneous image of the scientist. If motion picture portrayals are an indication of this image, the working scientist is a denizen of garrets and garages, an inept, absent-minded achiever in the laboratory but a nincompoop elsewhere. The newspapers too may be indicted for perpetuating a false image because they publicize the exceptional and extraordinary rather than the usual.

There have been scientists who have fit this generally false image. One J. Willard Gibbs, a foremost American physical chemist, whose phase rule is a bulwark of metallurgy, hardly ever spoke up at Yale University faculty meetings. He had no intimates and was alone practically all his life. Another scientist in this mold was eighteenth-century English chemist Henry Cavendish, who disliked and feared people. He was awkward, shy, homely, afraid of women, and seemingly indifferent to love and laughter. These are exceptions, but they have been presented as the rule. Their characteristics can be found in men and women in all walks of life, not in just one trade or profession.

Another popular image is that of the scientist who is usually a normal person but who has occasional periods when he is so immersed in his work that he forgets everything else. Anton von Leeuwenhoek, an early pioneer in science, became so absorbed with his lenses and what he saw through them that he neglected his family and was ridiculed by his neighbors, some of whom called him insane. The recipe for scientific

51

success given by early twentieth-century Russian physiologist Pavlov would engender such behavior. He advised: "Get up in the morning with your problem before you. Breakfast with it. Go to the laboratory with it. Eat your lunch with it. Take it home with you in the evening. Eat your dinner with it. Keep it before you after dinner. Go to bed with it on your mind. Dream about it." Again, this total involvement is not particularly characteristic of one field or another. Artists and writers, for example, can attest to being at one with a task as much as can scientists.

From the very beginning, scientists have been socially responsible people, participating in human events along with everyone else. Thales of Miletus, a leader of the Ionian school of Greek thought, was astute enough to corner the olive-press market. He predicted the solar eclipse of 585 B.C. He was among the first to state the principle of parsimony—that a maximum of phenomena should be explained by a minimum of hypotheses. Empedocles, the greatest democrat of antiquity, flourished about one hundred years after Thales, and he, like most of the early thinkers, was a universalist; as a physician he was in touch with the people. Archytas, a scientist and philosopher who was elected to be commander-in-chief of his city, Tarentum, several years in succession, lived about one hundred years later. Aristotle was active in the court of Philip of Macedon, father of Alexander the Great. A century after Aristotle, Strato was a staunch advocate of experiment and lived for some time in the king's palace at Alexandria.

The early Romans, renowned for their administrative skill, fostered engineers who were responsive to their communities. Strabo, the geographer, knew such influential politicians as the Prefect of the Egyptian province of the Roman Empire, Aelius Gallus. Some scholars believe that Strabo wrote his *Geography* for Queen Pythodoris of Pontus. At the age of twenty-eight, experimental physiologist Galen was appointed physician for the gladiators in Pergamon, a post equivalent to today's boxing-commission physician.

During the Middle Ages, many of the scientists of the expanding Islamic world were men of affairs. Avicenna, the outstanding physician, was not only the Sultan of Bukhara's physician, but also his general adviser. Ibn Khaldum of Tunis was not only a philosopher, but also a city planner whose *Muqaddimah* can be read today for its insight. The physician Averroes was summoned to Cordova because of his skill as a physician, but Caliph Yusuf found him immeasurably valuable as a general adviser and referred to him as "my wizard."

During the days of the Scholastics, nearly all of the famous were associated with the Church. They were socially responsible in the sense that they founded universities and hospitals. The number of hospitals in England grew about tenfold between the twelfth and thirteenth centuries. The Guild system of the Middle Ages did not neglect social responsibility. In nearly all guilds two of the "most loyal and befitting men" were elected to be overseers of work, to supervise and inspect the trade diligently, and to report every default without sparing anyone. The Scholastics may be considered forerunners of modern scientists in that they were thinkers, however often their basic assumptions were untenable. The craftsmen were predecessors of modern scientists by virtue of their work with things and instruments.

Each of the scientists who published important works around the middle of the sixteenth century held an official position. Nicholas Copernicus had a good post in Church administration; Andreas Vesalius served Charles V in Madrid; Ambroise Paré was a military surgeon for about thirty years, serving four successive French kings; Pope Paul III sought Jerome Cardan as an instructor in mathematics; Agricola, or Georg Bauer, the German Catholic physician who described mining practices, served as a burgomaster under a Protestant sovereign in a state where Protestants dominated.

An early Dutch investigator in mechanics, Simon Steven held an official government post; so did Dane Olaus Römer, who measured the speed of light through use of one of Jupiter's satellites. A mayor of Magdeburg, Germany, Otto von Guericke, did experiments dealing with air pressure and electricity. Martinius von Marum, an early pioneer in electricity, designer of the great electrostatic machine now in the Teyler Museum in Haarlem, Holland, was active in the city council there. The best publicist that early modern science had, Francis Bacon, was a member of Parliament and a prominent figure during the reigns of Elizabeth I and James I. Roger Boscovich, accomplished in many scientific fields, was at one time the ambassador of his native city Ragusa, then an independent republic in Yugoslavia, to the court of France. Astronomer Johannes Hevelius was a city councilor in Danzig.

Eighteenth-century scientists were also prominent as men of affairs. Mathematician and astronomer Pierre Simon, Marquis de la Place was either astute or diplomatic enough to survive several changes in the government of France, and he served in most of them; Napoleon, however, did not care for his introduction of the spirit of infinitesimals into government. A contemporary of Benjamin Thompson and Benjamin Franklin, English chemist Joseph Priestley sided with the American colonists and later the French revolutionists and was made a citizen of France. He emigrated from England to America after his liberal political views were opposed by a mob action. Chemist Antoine Lavoisier, founder of the present theory of combustion, was a member of the French tax-collecting body and was guillotined at Paris during the French revolution; his widow later married, and lived, unhappily, with Benjamin Thompson. Less than two weeks after Lavoisier's death, chemist Ignatius Martinovics, once in the employ of his ruler, the Emperor of Austria, was arrested for trying to spread the ideas of the French revolution; he was beheaded at Budapest in 1795. A scientist who fared better was chemist Antoine de Fourcroy. He was a member of the French convention during the revolution and later a counselor to Napoleon.

Science began to flourish during the nineteenth and early twentieth centuries, and those who cultivated science had less time to be men of affairs. Nonetheless their social responsibility continued unabated. Physiologist Rudolf Virchow was a candidate for political office in Germany. When English scientist Michael Faraday was invited to prepare poison gas for warfare purposes, he refused, claiming it to be inhuman and rejecting any connection with the activity. One of the 133 original members of the American Chemical Society, a physician who was head of the department of chemistry at Michigan Agricultural College, successfully campaigned for legislation against wallpapers printed with deadly arsenic pigments. Chemist Pierre E. M. Berthelot was France's Minister of Public Instruction during 1886–87 and the Minister of Foreign Affairs in 1895–96.

The social responsibility current among scientists has its roots in the past, but so too does the current social irresponsibility among scientists. The willingness of some scientists to serve any master is not new. Archimedes gave his skill to the tyrant of Syracuse, and Leonardo da Vinci marketed his abilities to a dictator. Perhaps the actual turning point for modern science, when social irresponsibility had a real birth, was Galileo's confrontation with the Church. At the time of his troubles, Galileo had lost some of the iconoclastic characteristics of his youth. By the time he was thirty-three, he could write astronomer Johannes Kepler: "I would dare publish my thoughts if there were many like you; but since there are not, I shall forebear." He was forty-five years old and possessed of more conservatism when his first book was published. He was sixty-eight and famous when the Inquisitors of Florence appeared at his house to serve him with a summons to be in Rome within thirty days. Friends advised him to escape; a sanctuary in Venice was offered. Instead, Galileo recanted his belief in the Copernican system. The tradition of science as a finder and supporter of truth became tarnished.

Galileo did what any man in his position would have done. But Galileo is not honored today as any man. He is worshipped by some as the father of modern science.

Giordano Bruno, a renegade Dominican and a humanist, was burned at the stake for his Copernican beliefs when Galileo was thirty-six years old. That punishment may well have had an adverse effect on Galileo's courage. Yet Galileo also had examples before him to bolster his courage. In January 1521, the Pope had declared Martin Luther excommunicated and had urged the Holy Roman Emperor, Charles V, to deny Luther the protection of the laws of the Empire. The Emperor had been willing to comply with the Pope's request, but the German princes had convinced him that Luther should be given a trial. Accordingly, Luther had been summoned before a great assembly, or diet, of princes and religious leaders at Worms. Here he had refused to retract anything he had written unless his statements could be shown to contradict the Bible. "It is neither right nor safe to act against conscience," said Luther. Since he had refused to recant, the Diet of Worms had proclaimed Luther to be a heretic and an outlaw. Luther had escaped death because the Elector of Saxony had protected him.

A similar challenge to science as a defender of truth was met more effectively by educator Thomas Henry Huxley during the nineteenth century. Huxley and others took up the argument against the fundamentalists, who opposed the organic evolution theory. Huxley and his friends presented facts and took part in public debate; they believed that in the marketplace of ideas the truth would be victorious over untruth.

In the 1950's, some American scientists did not follow the Huxley pattern. They made the mistake of appearing to be dogmatic, of refusing to recognize that ideas honestly presented must be challenged and honestly defeated rather than suppressed. The affair dealing with a book by Immanuel Velikovsky showed that some American scientists were ignorant of the meaning of communication in a free society. They need to understand this meaning if the problems science faces in communication and education are to be at least eased.

During the spring of 1950, *Harper's Magazine* printed an advance summary of Immanuel Velikovsky's *Worlds in Collision*. *Reader's Digest* published a shortened version of the book as well as a preface by a literary critic who claimed the volume could well be another *Origin of Species*. The book publisher listed the volume in its spring catalogue along with other new books. Scientists and others were offended by the category in which the book was placed in the catalogue—science. They disliked too the aggressive advertisement of it as well as the content which supported themes that modern science has denied.

Immanuel Velikovsky, born in Russia, earned an MD degree from the University of Moscow. During the late 1920's, he studied depth psychology with its pioneers in Vienna. He began to write a book he intended to call *Freud and His Heroes,* and in the course of preparation, he accosted the problem of dating the Exodus. He had the thought that the event could have been real. Following this line of thinking, he studied at the Columbia University library for months; he compared accounts of the Exodus in ancient documents.

Worlds in Collision maintains that the earth has been subject to many natural catastrophes that affected its rotation, orbit, and poles, and that many of these occurred in relatively recent times, being recorded in legends, history, and literature. One natural catastrophe he dated to the second millennium B.C. was the expulsion of Venus from Jupiter.

Velikovsky was not an astronomer nor a geologist, was not connected with universities or research institutions, and had not published in scientific journals. But to someone interested in ideas, the fact that an author lacks the usual academic credentials would seem immaterial.

Instead of offering a rebuttal to Velikovsky's interpretations of early writings, some scientists took it upon themselves to protest the publicity given the book and its suggested association with modern science. They hinted at a boycott of the publisher's textbook department. The organ of the American Association for the Advancement of Science, *Science*, made no mention of the book until it published a letter from Chester R. Longwell of the Department of Geology, Yale University, on April 13, 1951. Longwell quoted an editorial in the November 18, 1950, issue of the *Saturday Evening Post* entitled "The 1950 Silly Season Looks Unusually Silly," criticizing the "efforts of American scientists to suppress a book, *Worlds in Collision*." The *Post* editors claimed the scientists did succeed in forcing the original publishers to withdraw the book by threatening to boycott their textbook department. Another company did begin to publish the book. Longwell cited the editorial as silly and claimed that the original publisher's other authors had strongly protested.

During 1951 and 1952, some scientists wrote letters and articles, mostly in an unscientific vein, castigating the Velikovsky book. Occasionally a more rational analysis would appear, such as one by Laurence Lafleur in the November 1951 issue of the *Scientific Monthly*.

In 1953, the editors of *Science* published an article "Science, Poetry, and Politics" by a writer who had presented Velikovsky to *Harper's*. Eric Larrabee wrote about the barriers between the sciences and the humanities, a theme later publicized by Sir Charles Percy Snow. On August 7, 1953, the magazine published a letter from Professor J. P. Schafer of the Department of Geology, Brown University, identifying Larrabee as the *Harper's* editor responsible for giving space to Velikovsky and for generally defending him in spite of the attitude of the scientists. Larrabee's rebuttal was that the scientific community had not given serious consideration to Velikovsky's theses. He condemned practically all the criticism by scientists as "ill-considered, ill-tempered, and ill-informed . . . accompanied by a campaign of suppression, distortion, and intimidation."

Velikovsky was not shunned by the entire American science community. In 1953, he lectured at the graduate college of Princeton University. In 1954, he obtained Albert Einstein's help in suggesting to radio astronomers that they survey Jupiter to find its strong radio signals; the latter was found in April 1955. In 1956, Harry H. Hess, chairman of the Department of Geology at Princeton University, helped Velikovsky send a note to the American Committee for the International Geophysical Year to the effect that the earth's magnetic field might be stronger above the ionosphere and might have effects as far away as the moon; in December 1964, instruments on an artificial satellite did find evidence for this contention. In the December 21, 1962, issue of *Science,* a letter from Professor Valentine Bargmann of the Department of Physics at Princeton University and Professor Lloyd Motz of the Department of Astronomy at Columbia University noted that Velikovsky had predicted the high temperature of Venus measured by Mariner II on December 14, 1962, and that some other predictions of his had proved correct.

Whether Velikovsky is correct or incorrect, partly or wholly, his theories did receive much more diatribe than they did objective, rational analyses. The journals of science have not helped much; both *Science* and *Scientific American* rejected advertisements for the 1965 paperback edition of the Velikovsky book.

What are the chances for a recurrence of a Velikovsky situation or something similar? In 1964, a committee of the American Association for the Advancement of Science cited some recent examples of social irresponsibility among scientists. They claimed that military, commercial, political, and public pressures were eroding the integrity of science. The committee questioned the wisdom of very costly projects such as the one to put a man on the moon at the expense of needed research into the hazards of human survival. They charged that scientists had succumbed to military pressure in two space experiments where secrecy had prevented discussion which might have pointed out possible adverse consequences. One had been the Starfish project on July 9, 1962, in which a 1.4 megaton hydrogen bomb had been detonated 250 miles above Johnston Island in the Pacific Ocean. The intent had been to study the effect of the explosion on the Van Allen radiation belt. By 1965, the ionized space had not yet rid itself of the electrons released by the explosion. The second experiment they criticized was the West Ford project in 1963, in which an attempt had been made to establish an orbiting belt of fine copper wire around the earth to test a new type of communication system. The assurances that there would be no interference with radio astronomy studies proved false.

It may be that the social responsibility of scientists will always be menaced by the threat and practice of the antipode—social irresponsibility. Forces and personalities, controllable and not, will determine which one will be victorious. The battle may go one way or the other, depending upon the scientists. Whether a Leonardo da Vinci withholding the design of a submarine from evil men or a John Napier refusing to divulge his engines of destruction will stand as the dominating prototypes for scientists will be determined by the scientists themselves, beset as they are by mixed social, political, and economic forces.

The trend now seems to be positive. Despite the tiny percentage of scientists serving with school boards, civic committees, state legislatures, and Congress, scientists have been active. When atomic and hydrogen bomb testing was a menace, they were persuasive in calling for an end. American chemist Linus Pauling, a Nobel-prize winner in peace as well as in chemistry, was a leader in this movement. In 1958 and in 1962, he and his cohorts, including ten Nobel laureates, sought an injunction in the United States Federal Court and with the Procurator General of the U.S.S.R. to stop the testing. World leaders pay attention to the Pugwash deliberations of scientists; now called Conferences on Science and World Affairs and attended by invited scientists of many nations, the Pugwash meetings have spearheaded advances in relieving international tensions.

Scientists have been active within their communities. In St. Louis, their concern made them distribute vital nuclear information to the public; this action has developed into a national movement dealing with science and public affairs and has led to the issuance of a magazine first called *Science and Citizen* and now called *Environment*. In 1959, men from the Schenectady, New York, area began Volunteers for International Technical Assistance (VITA) and now have more than one thousand members helping to solve technical problems arising in the underdeveloped countries. The Western Montana Scientists' Committee for Public Information, centered at the University of Montana at Missoula, never had more than fifteen members but was able to work successfully for many antipollution measures. In 1969, an association of biologists in Maine launched a boycott of the products of a potato processing firm accused of environmental pollution. The Society for Social Responsibility in Science, founded in 1952, has more than one thousand members, many of them in foreign countries, and strives for constructive use of scientific and ethical criteria in choice of work. The unaffiliated British Society for Social Responsibility in Science was founded in 1968.

Social responsibility is now wholeheartedly accepted by the leaders in science. The editor of the conservative *Chemical and Engineering News* wrote in the January 8, 1968, issue: "The scientist who denies he should be actively concerned over the use of scientific knowledge he develops, yet expects to be supported morally and financially by a society increasingly concerned with his influence on it, needs to change his philosophy." Dr. Max Tishler, president of the Merck, Sharp & Dohme Research Laboratories, said before the Royal Society of Medicine in London on January 8, 1968: "Science was originally responsible only to a few. It now has to adapt itself to new responsibilities to the many. This means that scientists, in medicine and other fields, must get into public affairs because public affairs are getting into science." In these days scientists have become involved in public affairs from the antiballistic missile and environmental pollution to the inner city and education.

SELECTED REFERENCES

De Grazia, A. *The Velikovsky Affair*. New Hyde Park, New York: University Books, 1966.

Goran, M. "Science Leaves the Ivory Tower," *The Humanist*, 3 (Autumn, 1943), 100–02.

Larrabee, E. "Scientists in Collision," *Harper's Magazine* (August, 1963).

Price, D. K. *The Scientific Estate*. Cambridge: Harvard University Press, 1965.

Russell, B. "The Social Responsibilities of Scientists," *Science*, 131 (February 12, 1960), 391–92.

Seaborg, G. T. "Scientists Shed Wicked Wizard Tag," *Chemical and Engineering News*, 42 (December 21, 1964), 60–62.

Smith, A. K. *A Peril and a Hope: The Scientist's Movement in the U.S. 1945–47*. Chicago: University of Chicago Press, 1965.

6

TRESPASSING

In a free community it is the inalienable right of every citizen to have and express an opinion about any subject in the public domain. Indeed, the citizen must have opinions in the area of public affairs if a democracy is to function effectively. In fulfilling this responsibility, every person runs the risk of making an ass of himself, particularly if he is not informed or if his thought processes are poor. The scientist as a citizen is no exception. He has opinions and he expresses them; sometimes they are foolish and mistaken. Unfortunately, the average nonscientist views the scientist's stand on public affairs as that of a thinker and a scientist. In yesterday's nonspecialist world, this view was tenable. The scientist was more or less well informed on all affairs. In today's working universe of subdivided and minute areas of authority, the scientist can no longer be respected for his ideas in all fields. Physicist Eugene P. Wigner, at the press conference at Princeton University following his award of the Nobel prize said: "It does not make me a person of wisdom. . . . It is a great danger if statements of scientists outside their field are taken too seriously."

In the November 22, 1965 issue of *U.S. News and World Report,* American Nobel-prize winner in physics William Shockley stated his opinions on the biological and sociological question of whether the quality of United States population is declining because of hereditary factors. At the annual meeting of the National Academy of Sciences in Washington, D.C., April 24–26, 1967, Shockley continued his comments on the social sciences. He wanted a study made to test the theory that genetic, rather than environmental factors, are responsible for the poor performance of some Negroes in competitive situations. The respected English science weekly *Nature* editorialized: "Professor Shockley has not merely been guilty of bad taste. . . . [A]lthough there are some who will say that Professor Shockley should have stuck to his last, the real lesson is that he should have done his homework much more carefully." Later in the year Shockley suggested to about four hundred listeners at McMaster University in Hamilton, Ontario, that the population problem could be eased by temporarily sterilizing all young girls.

Along with others, scientists have expressed strong prejudices in the area of race. When modern science was largely the monopoly of western Europe, these prejudices were not significant; the opinions were not widely distributed. With the growth of new nations and the development of all peoples in the twentieth century, the prejudices have had wider circulation. German scientists were among the first in the field of amateur racism. J. C. F. Zöllner, a professor of physics at Leipzig University, was a notorious anti-Semite during the late nineteenth century. (He was also involved in spiritualism, conducting experiments with a fraudulent American medium.) Another trespasser on the subject of race was engineer Theodor Fritsch in the same city at about the same time; Fritsch, a voluminous writer, became a Nazi Reichstag deputy before he died in September 1933. Two German Nobel-prize winners in physics, Philipp Lenard and Johannes Stark, were the favorites of the Nazis during their rise to power, and both had expressed their anti-Jewish prejudices even before the Munich beer hall putsch of the early 1920's. When scientists outside Germany protested the treatment of German-Jewish scientists, both Stark and Lenard rushed to the defense of the Nazis. Stark sent several letters of explanation to *Nature*. He denied that large numbers of German-Jewish scientists had been exiled, and he termed the Nazi race policy a good one for Germany.

The most recent trespassing by scientists into the race problem was triggered by the civil rights struggles of American blacks. In 1963, *Science* printed letters from southern white racists, with such specialties as anatomy, and rebuttals by other scientists not so inclined. On October 16, 1964, *Science* published an article entitled "Racial Differences and the Future" by physiologist Dwight J. Ingle of the University of Chicago. The reaction was immediate and overwhelming. The journal had to devote several pages of two issues in December 1964 to printing the protesting letters from sociologists and anthropologists. One correspondent scolded the editors for publishing the article during a national election when the civil rights of blacks was an issue for many. Ingle answered not only the critics, but also black Congressman Adam Clayton Powell, who expressed his opinions in the issue of January 1, 1965. Ingle also published "The Biological Future of Man" in the Spring 1966 issue of the University of Chicago alumni magazine.

More trespassing must necessarily occur as scientists become involved in the great social issues of the day. When seven hundred scientists gathered in New York on June 15–17, 1962, for the Congress of Scientists on Survival, the result was a potpourri of opinion on many topics. A prominent physicist spoke about the diversion of funds from various projects to a campaign for world peace; a sociologist commented on the effects of blasts of various kinds on concrete; some anthropologists talked about fire storms. At the September 1964 Conference on the Impact of Science and Technology on Contemporary Society, held in Yugoslavia, the scientists talked about science in politics and about laws of economic growth as well as about more clasical scientific topics. Letters to the editor and printed addresses in American science journals have comments on such topics as Keynesian economics and the family as a transmitter of culture. The President's Science Advisory Commission has been reported to discuss not only science problems, but also such issues as the quality of American life and the upgrading of elementary education.

Trespassing has overtones of being one of the solutions to the problems of science education because trespassing by scientists, as well as by non-scientists, into fields other than their own, has been very fruitful. Many pioneers of modern science who had the benefits of advanced schooling were trained in fields other than those in which they won recognition. Agricola, author of *De Re Metallica,* published in the sixteenth century, was a physician; so was the expert on magnetism who lived a little later in England, William Gilbert. The outstanding geologist of the nineteenth century, Sir Charles Lyell, was first a lawyer; so was Edward Lartet, one of the founders of physical anthropology, as was Edwin Hubble, early twentieth-century American specialist in galaxies. The renowned historian of science George Sarton earned his Ph.D. degree in mathematics, and the present-day expert on Galileo, Stillman Drake, was a banker before joining the faculty of the University of Toronto.

Many pioneers in science missed the benefits of formal higher education, and so they may be considered trespassers into the arena of the learned. Blaise Pascal in the seventeenth century and Benjamin Thompson in the eighteenth century had no university training; neither did Humphry Davy or Michael Faraday in the nineteenth century. William "Strata" Smith was an untutored canal builder who in the early nineteenth century established the relationship between layers of rock and the fossils they contain.

The self-taught, regardless of their formal education pattern, are numerous in science even now. Among the great twentieth-century scientists who considered themselves self-taught, despite their lengthy training in science, were organic chemist Adolf von Baeyer, chemist Fritz Haber, and physicist Enrico Fermi. They would thus be in the same self-taught category as other prime innovators of years ago such as George Boole, sometimes called the most underrated mathematician of the nineteenth century; his *An Investigation into the Laws of Thought* was published in 1854.

The fact that there are so many self-taught scientists (whether they assign themselves to this category or are assigned to it) indicates that there are people in science who do not hesitate to move into problems seemingly removed from their experience or training. This kind of trespassing has resulted in the establishment of not only new ideas, but also new subjects; such borderline subjects as biophysics and molecular genetics needed and found such innovators.

Trespassing of this kind may perhaps be done by any scientist. During World War II it was not uncommon for an American biologist to be successful with radar problems or a physicist to be an expert designer. On the other hand the lurking danger of failure is omnipresent. Early in the twentieth century, medical men described certain artifacts found in southwestern United States as ancient therapeutic corsets. Later study by archeologists corrected this view; the therapeutic corsets proved to be cradleboard hoods.

Other worthwhile trespassing by scientists may underscore what educators and some scientists have been claiming for the ways of science: they need to be used in all walks of life. Philosopher John Dewey said that the advance of mankind depends upon the widening spread and deepening hold of scientific procedures. Depth psychologist Sigmund Freud insisted that if there were any hope for mankind, it would lie in the scientific method. When scientists become successful in nonscience areas, it may be that their training in scientific thinking was an important asset. The Soviets choose a large percentage of their managerial class from those trained in science. In the United States, a 1958 investigation revealed that among sixtythree federal government bureau chiefs, nine had advanced degrees in the natural sciences and seventeen others had some kind of engineering or technical background. This kind of trespassing is safeguarded by the critical faculties of others, the performance of the ex-scientists, and the retraining they necessarily undergo to be effective in their new jobs. This kind of retreading is more widespread that the sample figures indicate. Economic rewards in business, government, and schools are greater to the management class, a fact many scientists first realize when they begin to work for a living. Some change their professional goals accordingly. Evidence of this realization is given by the large number of engineers and scientists who apply to graduate schools of business to try to earn a master of business administration degree. To thwart this trend, a few forward-looking organizations have installed pay scales for scientists comparable to those for administrators.

SELECTED REFERENCES

Goran, M. "Scientists Are Also Writers," *Journal of Chemical Education,* 39 (September, 1962), 479–80.

————. "Science as Art," *Journal of General Education,* 18 (January, 1967), 281–88.

Letters to the Editor, "Race, Science, and Social Policy," *Science,* 146 (December 11 and 18, 1964), 1415–18, 1526–30.

Stark, J. "Letter to the Editor," *Nature,* 133 (February, 1934), 290; (April, 1934), 615–16. "Replies by A. V. Hill," *Nature,* 133 (February, 1934), 291; (April, 1934), 616.

7
HONESTY

If transfer of training does occur, one aspect of science has great importance not only for its problems of communication and education but also for some of the difficulties of mankind. This characteristic of science has been described many times by articulate scientists and occasionally by others. American philosopher Sidney Hook said it with: "The cardinal sins in natural science are to cook the facts, to misreport, and to suppress evidence. All the players have clean hands; the rules of the game have become second nature to the participants."

When individual cases are considered, the meaning of honesty can become a vexing issue. When Polonius in Shakespeare's *Hamlet* advises his son Laertes that "this above all, to thine own self be true," he acknowledges a kind of honesty, but what person knows himself completely?

If honesty is taken to mean the non-withholding of information, then dishonesty is rife in applied science and engineering. Many business and industrial organizations depend upon some well-kept secret and will fight to protect it. Scientists and engineers who enter the employ of such companies often sign statements that they will not divulge information; they sometimes must promise not to enter the employ of a competing company.

In 1964, the United States Court of Appeals ruled against a young chemist named Wohlgemuth, who was hired by the International Latex Company after six years of work in the spacesuit technology division of the B. F. Goodrich Company. The court cited that the young man had no moral compunctions against disclosing Goodrich secrets when he left that company's employ. Twice in their decision they commented on Wohlgemuth's attitude, quoting him as saying that ". . . loyalty and ethics had their price"; insofar as he was concerned, International Latex was paying the price. He said that "once he was a member of the Latex team, he would expect to use all of the knowledge that he had to their benefit." The Court of Appeals concluded that "public policy demands commercial morality and courts of equity are empowered to enforce it by enjoining an improper disclosure of trade secrets." In this case, the lower courts commented on what was called the wrongful conduct of the International Latex Company in luring Wohlgemuth "to induce, if possible, the defendant in this case to give them the benefit of every kind of information he had."

Trade secrets may be carefully kept and protected by law, but when they are such that will give others profit or power, they are not kept very long. The secrets of the obstetrical forceps were once diligently guarded; so were some lens construction details. As recently as 1945, there were intelligent people in the United States and England who thought that the construction details of the atomic bomb could also be kept secret.

If honesty is equated with giving all information freely, then dishonesty has sometimes occurred in pure science. There was an attempt to withhold data about the first chemical element discovered in modern times, phosphorus. As states came to finance and control pure science, particularly during and after World War II, information normally distributed was not released. At times, scientists chafed at the bureaucrats who would class the periodic table of chemical elements, known since the middle of the nineteenth century, as restricted information.

The lack of a complete and exhaustive report of scientific research in today's journals is not a form of this kind of withholding. Despite the thousands of scientific journals available, there is not enough space to print a record of all the false starts, the hypotheses, and the insights of the investigators. Because of the necessity of economy, scientific reports are terse and devoid of appealing style.

It has been said that a great deal of neglecting to get consent, comparable to withholding of information, occurs in the medical sciences. An indignant member of a board of directors of a New York hospital revealed that certain doctors had injected live cancer cells into old, helpless patients against the protests of their physicians; neither the patients nor the physicians had consented to such an experiment. In January 1966, the Regents of the University of the State of New York, acting under their responsibility for licensing the medical profession, found the doctors involved in the experiment guilty of "unprofessional conduct" and of "fraud and deceit in the practice of medicine."

In 1966, Dr. Henry K. Beecher of the Harvard Medical School described twenty-two cases in the *New England Journal of Medicine* of recent physiological experiments done without the subjects' consent; twenty-eight other cases were not presented because of lack of journal space. In one of the cases described, penicillin was withheld from a control group of servicemen with streptococcal infections even though the antibiotic is known to be a preventive for rheumatic fever; several men thereupon developed the disease. In 1967, English physician Maurice Pappworth published his *Human Guinea Pigs,* wherein he claimed that children, pregnant women, mental defectives, and old people were experimented with in British hospitals without either their consent or knowledge.

The public interest in such activity by scientists may have been a factor in the grant made by the U.S. Public Health Association to the American Academy of Arts and Sciences for an inquiry into the moral and ethical bases for research involving human subjects. Eight U.S. medical organizations have already endorsed the ethical principles approved in 1964 in the Declaration of Helsinki by the World Medical Association concerning human experimentation. The Declaration of Helsinki states that if a patient is legally incompetent, the consent of the legal guardian should be obtained by the researcher; at any time during the course of the research, the subject or his guardian should be free to withdraw permission for research.

The distortion of facts is a form of dishonesty only if it is done knowingly. Scientists must continually interpret and reinterpret facts, and sometimes distortion can creep in. Very early scientists interpreted the daily turn of the skies as simply that— a change—while modern scientists view the diurnal change as a reflection of earth rotation. Early chemists believed burning to be a release of a substance called phlogiston, while modern scientists know burning is a process of combination with oxygen. The early scientists invented negative weight to keep alive the concept of phlogiston. When confronted with the fact that weights of substances before chemical reaction equal weights of substances and all gaseous products after chemical reaction, they said phlogiston had a negative weight. At one time, the early investigator in chemistry Henry Cavendish thought he had isolated phlogiston.

A difference of opinion, vigorous or not, in a meaningful area or not, certainly does not reflect dishonesty, despite the fact that the daily press sometimes infers overtones of dishonesty. Early in November 1964, as a result of the presentation of two almost diametrically opposed conclusions by the Atomic Energy Commission agencies, the Pacific Gas and Electric Company withdrew its application to construct the Bodega Bay, California, atomic power plant at a site close to an established earthquake fault. The Atomic Energy Commission Advisory Committee on Reactor Safeguards confirmed the safety of the plant site; the regulatory staff of the Atomic Energy Commission decided that the site was not safe. Both groups had the advice of eminent authorities.

The falsification of facts is distinctly dishonest, and in the annals of science there is one celebrated case and several minor ones. Among the latter cases is the one of mathematician Jean Bernoulli, who at the beginning of the eighteenth century was accused of dishonesty at Groningen University and suspected of cheating at a prize competition at Basel. Pascual Jordan, the twentieth-century German physicist, knew of only three cases of scientists deliberately submitting false data. Recently, Mendel and Darwin have been accused of this practice, but hardly any evidence is available that supports such an accusation.

This type of dishonesty does not occur too often in science, despite the lucid presentation in C. P. Snow's novel *The Affair*. The frequency is rare because of built-in safeguards in the procedure of science. Data recorded in journals is subject to verification. Such replication is the check that prevents false records. If the result cannot be duplicated, the original research is held in question.

The major dishonesty case arose in the science of biology

early in the twentieth century. After the theory of organic evolution had been successfully defended against scientific critics
and fundamentalists, it was necessary to determine the usefulness of Darwin's suggested mechanism for evolution. He used
the term natural selection to describe the processes in nature
whereby one species gradually became another. At the turn of
the century, new mechanisms were suggested and an old one
was revived. One new idea was that natural events produced
changes in the germplasm, the nucleus of the cell, and these
changes were transmitted to succeeding generations; such muta-
tions abetted by natural selection were championed by many
biologists during the early twentieth century. The competing
idea was one defended by Lamarck; it saw the progress of organic evolution as being caused by the inheritance of acquired
characteristics. The argument was that an ancestral giraffe had
acquired a long neck by use and that this long-necked giraffe
had passed on the characteristic to his progeny. Inheritance of
acquired characteristics suited the philosophical base of the
communist world. The communists looked with favor upon
this idea because it seemed to vindicate their theme that environment could successfully and permanently alter a species,
and presumably man.

Vienna-born Paul Kammerer was among the scientists who
investigated inheritance of acquired characteristics. His interest in the subject came late in life, though he had been involved in allied research for most of his career. He was born in
1880; the son of a manufacturer of scientific instruments. Kammerer studied at the University of Vienna, and he remained in
the city until 1923, working all that time at one position with
the state-sponsored biology research organization.

Kammerer's research dealt with the experimental modi-
fication of animals; he won the 1909 Sömmering medal
awarded by the Seuckenbergische Naturforschende Gesellschaft
in Frankfurt-on-Main for some of his first work. He showed
that the reproductive process in amphibians and reptiles was
altered by a change in temperature and moisture. He investigated the effect of background color upon organisms, and he
found one species of salamander capable of changing its color
according to the color of light reaching its eyes.

Toward the end of World War I, he began to work with alytes, midwife toads. The male of the species carries the female's eggs on his legs until they hatch under water. He raised one group of these toads out of water so that the eggs were hatched in air; he maintained another comparable group so that the eggs were hatched in water. His experiment suggested that the latter after a few generations developed new features called nuptial pads, small horny pads on the thumbs, useful for holding on to mates in water.

Kammerer's report appeared in 1919 in the *Archiv für Entwicklungsmechanik,* where practically all his research results were published. The interest of biologists in the work became widespread. In 1923, Kammerer, although only forty-three years old, applied for his pension and began lecturing throughout Europe; he also made two trips to the United States. He published a book in Germany called *Neuvererbung* (Stuttgart-Heilbronn: W. Seifert-Verlag, 1925) and one in the United States, *The Inheritance of Acquired Characteristics* (New York: Boni and Liveright, 1924). In 1925, he was given a professorship at the State University in Moscow and was asked to supervise the building of a laboratory in the biological department of the Academy of Sciences.

Kammerer was acclaimed among some biologists. Dr. Thornley Garden, professor of zoology at Cambridge University, said, "Kammerer begins where Darwin left off." Dr. G. H. F. Nuttall, professor of biology at Cambridge University, claimed, "He has made perhaps the greatest biological discovery of the century." More amazing than the testimonials was the fact that no one attempted duplication of such an important discovery. When, for example, news that certain uranium atoms bombarded with slow neutrons released tremendous energy reached the United States, concerned scientists quickly duplicated the experiment to verify the result. The latter was highly important and had to be substantiated. A similar process did not ensue with Kammerer's work.

By 1926, only a single specimen was available to support the claims of Kammerer. It had been brought by him to England in support of his lecture statements. In 1926, the specimen was at the Biologischen Versuchanstalt in Vienna, Kammerer's former place of employment. A young scientist, Dr. G. K. Noble of the American Museum of Natural History in New York, visited the Vienna Institute in 1926 to examine Kammerer's work. His preliminary examination of the specimen revealed enough unexpected features to have him undertake a complete examination. He found that India ink had been injected into the specimen, and the usually black little pads were smooth rather than spiny. Noble concluded in a report to *Nature*: "It has been established beyond a shadow of a doubt that the only one of Kammerer's modified specimens of alytes now in existence lacks all traces of nuptial pads."

Noble's work, although critical, did not give a complete answer to the issues raised. Kammerer had said in 1923 that "dozens of scientific men have seen the pads and are now convinced." Furthermore, a photograph of the specimen had been seen by many professionals in England who acknowledged Kammerer's claim. It was necessary to explain away these events.

Noble's letter was printed in the August 7, 1926, issue of *Nature*. The very next letter in the same issue was from Dr. Hans Przibram of the Biologischen Versuchanstalt in Vienna. He agreed with Noble's conclusion but proceeded to explain it away: It could be that repeated handling and shaking of the specimen was the cause of the loss of the nuptial pads. He quoted the opinion of experts who had seen photographs of the specimen before it had gone to England. One expert photographer said: "I have examined carefully the negative of the alytes, and while it appears to have been intensified and a string across the background retouched, there are no signs of any retouching or interference with the image of the specimen itself, or of that part of the background with which it is in contact." He quoted a scientist at Trinity College, Cambridge, who wrote: "In April 1923, I had many opportunities of examining the specimen, and was always able to see the spines, whether by means of a lens or a dissecting microscope, exactly as in the photograph in question." He listed English zoologists who had examined the specimen and presumably had seen the nuptial pads.

In the October 16, 1926, issue of *Nature,* Dr. Hans Przibram wrote to report that Dr. Paul Kammerer had shot himself when on the Hochschneeberg, near Vienna, on September 23, 1926. "In a letter received after his death," wrote Dr. Przibram, "he emphasizes that he never committed the scientific tricks hinted at by some of his critics . . . [he] alleges that someone might have manipulated it; he does not allude to a suspicion who this might have been."

The Presidium of the Moscow Academy also received a letter from Kammerer. He wrote: "After having read the attack, I went to the Biological Experimental Institute for the purpose of looking over the object in question. I found the statements of Dr. Noble completely verified. Indeed, there were still other objects (blackened salamanders) upon which my results had plainly been "improved" post mortem with India ink. Who besides myself had any interest in perpetuating such falsifications can only be very dimly suspected. But it is certain that practically my whole life's work is placed in doubt by it.

"On the basis of this state of affairs I dare not, although I myself have no part in these falsifications of my proof specimen, any longer consider myself the proper man to accept your call. I see that I am also not in a position to endure this wrecking of my life's work, and I hope that I shall gather enough courage and strength to put an end to my wretched life tomorrow."

On October 30, 1926, *Nature* published an obituary for Paul Kammerer. In the description of his previous accomplishments were included the names of some scientists who had verified the doubted work. At the end was the statement: "Kammerer's last paper on the origin of the island races of Lacerta in the Adriatic (1926) is one of the finest contributions to the theory of evolution which has appeared since Darwin."

Przibram again tried to establish the integrity of Kammerer in a letter to *Nature,* April 30, 1927. In it he absolved one of Kammerer's co-workers from the misdeed and gave several evidences that the injection of India ink had been done after Kammerer's work. He alluded to "the new experimental evidence which Kammerer's collaborators in Moscow are trying to get."

Even if Kammerer or one or more of his co-workers were extra diligent or dishonest, the problem of the other scientists who testified to the work remains. If the specimen they saw were a fake, their reaction could perhaps be explained on the basis of mind sets. An individual with a mind set can see "4 percent bonds" as "4 perfect blondes," or "God hates a quitter" as "Good hats a quarter." How else can the same object be seen differently? Equally competent astronomers will look directly at Mars or at photographs of Mars and be divided on what was seen; some will see a fine network of lines and others will not notice any such lines.

If the specimen Noble examined was the only one that had been tampered with and Kammerer's results had been accurate, then his reaction of committing suicide was strange. He could have demonstrated the truth of his conclusions by simply repeating his work. In his letter to Przibram, he deemed the rest of his life too short to be able to take up again the same experiments; he said he was too weary. He was then forty-six years old.

There have been other catastrophic experiment failures in the history of science, but the reaction of the investigators was not suicide. One case in Germany provoked overwhelming embarrassment. At the end of World War I, defeated Germany was faced with the problem of paying the reparations demanded by the Allies. Fritz Haber, developer of the nitrogen-fixation process that gave the Kaiser much-needed fertilizers and the base for explosives, had the idea of sweeping the oceans for gold. He gathered financial support, outfitted ships, tested extraction means, and made analyses. In the end he had to drop the project; the values of the amount of gold in ocean waters that he had assumed from preliminary records of previous investigators, as well as from his own preliminary analyses, were too large. His research should have started with an exhaustive sampling of ocean waters. Yet in the course of the work, a very sensitive test for gold was developed. After the project to find gold in ocean water had been dropped, another scientist came to Haber to demonstrate the conversion of mercury into gold. (This was in the days before the development of modern nuclear equipment which has made possible such a conversion, albeit at very great cost—indeed, more than the worth of the gold produced.) Haber and assembled scientists watched the demonstration and Haber was greatly impressed. "Gentlemen, we were here," he said at the conclusion. He thought that the demonstration would be an epoch-making one and that the scientists were privileged to be onlookers. In due time others in the laboratory showed that the small amount of gold seemingly produced had come from such sources as the demonstrator's gold-plated spectacles, that no element conversion had occurred. Haber was forced into the embarrassing position of repudiating his prior comments and he did so with force, excoriating the demonstrator who had honestly believed he had accomplished the change.

SELECTED REFERENCES

Ley, W. *Salamanders and Other Wonders*. New York: Viking, 1955.
Noble, G. K. "Kammerer's Alytes," *Nature,* 118 (August 7, 1926), 209–10.
"Obituary of Paul Kammerer," *Nature,* 118 (October 30, 1926), 635–36.
Silverberg, R. *Scientists and Scoundrels*. New York: Crowell, 1965.
"Trade Secrets: The Technical Man in Legal Land," *Chemical and Engineering News,* 43 (January 18, 1965), 81–94.

8
THE DANGER OF DOGMA

Science's problems of communication and education are made difficult by the prevalence of dogmatists. The nonscientist is taken aback by the arrogance and dedication of those who wish to apply every aspect of science elsewhere. The science student somehow comes to believe that science courses in schools are simply indoctrination.

Dogma in science has a long history. One of the first pioneers to bear its brunt was Andreas Vesalius. He was born in Brussels, Belgium, on December 31, 1514. His great grandfather had been a physician to Maxmilian I of Burgundy; his grandfather had written a commentary on a famous textbook by Rhazes, the Arab; his father was an apothecary, a drug dispenser, to Emperor Charles V of the Holy Roman Empire. After a lengthy education Vesalius was appointed a professor at the famous University of Padua. Here he was diligent in conducting dissections to establish human anatomy; the work by the early Roman experimental physiologist Galen was taken as the authority, and the efforts of Vesalius were in essence to substantiate what Galen had reported more than one thousand years before. Vesalius' dissections showed that the actual course of the azygos vein was entirely at odds with Galen's description; consequently he published a small book called *Letters on Vein Cutting*.

In 1540, Vesalius became a member of a team intending to translate Galen's works into Latin. At this time Vesalius was lecturing at Bologna, and as part of his course of lectures, he rebuilt the skeleton of a man and that of a monkey. It was at this point that he began to realize that Galen's works described monkeys, not men. He wrote: "I could not get over wondering at my own stupidity and overconfidence in Galen and the writings of the other anatomists."

It took him two years to write his *De Humani Corporis Fabrica*; it was published in June 1543, when he was twenty-eight years old. Soon after it appeared, dogmatists came to the fore. One of Vesalius' former teachers at Paris, Jacob Sylvius, called him wholly arrogant and wholly ignorant, a madman and an unprincipled upstart. It may be that Sylvius really wanted more credit—in the preface to *Fabrica,* he is mentioned as "the never-to-be-sufficiently-praised Jacobus Sylvius"—but his diatribe, aside from the calumny, was a passionate defense of Galen.

Issac Newton presents a case where a dogmatic approach toward one's work, together with the acceptance of that work by others as dogma, can hinder not only scientific progress, but also the philosophy of science. Newton was a genius who advanced science tremendously, but his attitudes were not always in the best interests of science.

There is no record of accomplishment in the family of this genius. His people had been farmers. His father had died before Isaac was born on Christmas day in 1642, and his mother soon made a second marriage for convenience. It may be only coincidence, but Newton's being raised away from his mother's house because of the remarriage parallels the case of the Italian-American genius Enrico Fermi, who was so frail as an infant that he was sent to spend a couple of years in the country away from his family. Two outstanding physicists of the nineteenth century, both Scotch, James Clerk Maxwell and Lord Kelvin, lost their mothers early in life and were raised by their fathers. Many other pioneers of science had but one parent during infancy or childhood.

There is no record of Newton showing his talents while he was a boy at school. Early signs of genius are usually not as discernible to onlookers as they are to biographers. Newton did display a liking for mechanical things as well as for books. From his own records and notes, Newton reveals himself during his boyhood years as having a quick temper.

In 1661, Newton went to Cambridge University and remained

there for forty years, except for relatively short periods. When he became a professor, his work was not taxing and it allowed much free time. He delivered one lecture a week, one-half hour long, from October through December. He had the kind of position termed ideal by the present-day research-oriented American professoriat.

In one of his stays away from Cambridge University, he was director of the Mint in London for seven years. Here he had a practical accomplishment to add to his long list of scientific ones. Before he became director of the Mint, the average value of a coin had been reduced by the habit of some people to clip metal off coins. These coins made transactions difficult because some people wanted to exchange on the basis of weight and others on the basis of value indicated. Newton and his friend Lord Halifax succeeded in recalling old currency and substituting circularly milled coins of full weight.

For a time Newton was a member of Parliament, but he never spoke in debates. He was elected by his colleagues at the university who appreciated his defense of their rights. James II had attempted to turn England into a Roman Catholic country, and Newton among others had resisted.

Newton's achievements in science were gigantic. As a theoretician, he was an independent co-founder of the important branch of mathematics called calculus. His *Principia* established mechanics on a firm basis and led to its universal acceptance; his three laws of motion and law of universal gravitation still offer excellent explanations of terrestrial and astronomical phenomena. As an experimenter, he dabbled in alchemy. He made excellent observations in optics. As a reporter of science, he ran the gamut from the cold, terse style of his *Principia* to the warm, friendly descriptions in his *Opticks*. He wrote hundreds of letters and left scores of unpublished notes and manuscripts. He had the habit of annotating what he read and often copied a great deal of it. As a criminal investigator (part of his duties at the Mint), he was effective in finding counterfeiters; an organized police system was not to come until after his tenure. He made minor contributions to theology, where he was very much involved; he thought that the great usefulness of science was in its support of religion. With such a record of accomplishment, any mistake he made would seem to be forgivable.

Even his personality can be overlooked on account of his contributions to science. He was a cold and aloof person, given to sudden bursts of ill temper and seemingly forever locked in controversy with others. One recent critic accused him of being a neurotic who was afraid to show his thoughts and discoveries. His successor at Cambridge University said Newton was "of the most fearful, cautious, and suspicious temper that I ever knew." He had nothing to do with the opposite sex and once accused philosopher John Locke of trying to embroil him with women. Yet he seemed to condone the arrangement between his niece and Charles Montague, later Lord Halifax.

Newton's dogmatic attitude showed first in his avoidance of scientific debate. Scientific facts and theories are public ones, open to the scrutiny of all and available to all. If Newton's attitude toward communicating scientific ideas were adopted, the openness of science and its free flow of information would come to an end. Perhaps if his first contribution to science had been received with adulation and accolade, he might have been less touchy about publishing his work. He had to be pressured to communicate practically all his research. His first report to the Royal Society was no more adversely criticized than are the scientific reports of today; indeed, he was given recognition for the work. But the small amount of scientific doubt raised about his work was enough to cause him regrets about having informed others of his research. Many times thereafter he protested that he would abandon science or that he would withhold his work from publication during his lifetime.

He had scientific arguments with Robert Hooke, John Flamsteed, and Gottfried Leibniz, and these arguments he interpreted as attacks on his personal integrity rather than scholarly differences of opinion. He became morbidly sensitive to any opposition. If an explanation of his ideas did not meet the needs of his correspondents, he simply stopped writing. His twentieth-century American biographer Louis T. More wrote that "if he was further opposed, he became an implacable and, even at times, an unscrupulous antagonist."

Newton's critics, the men who presumably prompted him to

acquire his defensiveness, were intelligent men of the day and
in at least two cases were brilliant scientists in their own right.
Robert Hooke, author of *Micrographia,* a book which to some
is as remarkable as Newton's *Principia,* was one. In Hooke's
work there is mention of the cell idea that was later so useful
to biology, and there are also superb drawings of many organisms. Some scholars see Hooke as antedating Newton in arriv-
ing at a universal law of gravitation. Another was Christian
Huyghens, often called the Dutch Archimedes, the author of a
theory about light competitive to Newton's.

Newton's secretiveness may have stemmed not only from his
fundamental character and from the scholarly criticism he had
received, but also from the fact that his first work was rejected
for publication. His edition of Kiukhuysen's *Algebra* had been
declined by the university press for the reason that they had
other work on hand; Newton was then thirty-three years old. It
is also possible that he looked at science with the attitude of a
theologian: Nature was a mechanism whose secrets were to be
found once and for all time, and therefore the results were not
subject to debate and change.

Newton's dogmatism is also reflected by his use of the phrase
hypotheses non fingo—I frame no hypotheses. He certainly used
hypotheses in the modern sense, and early in his career he had a
forward-looking view of scientific procedure. When he was
thirty, he wrote to the German-born secretary of the Royal
Society Henry Oldenburg: "You know the proper method for
enquiring after the properties of things is to deduce them from
experiments." At about the same time, he wrote Father Ignatius
Pardies, a professor of natural philosophy in the College of Cler-
mont at Paris, a clear view of the nature of hypotheses: "For
the best and safest method of philosophizing seems to be, first to
enquire diligently into the properties of things, and of establishing those properties by experiments, and then to proceed more
slowly to hypotheses for the explanation of them. For hypotheses
should be subservient only in explaining the properties of
things, but not assumed in determining them; unless so far as
they may furnish experiments."

Newton's *Principia* was published when he was forty-five years
old. Toward the end of the book, he mentions the cause of grav-

itation and writes in an entirely different vein than he did to Father Pardies: "Here I have not been able to discover the cause of those properties of gravity from phenomena, and I frame no hypotheses, *hypotheses non fingo,* and hypotheses whether metaphysical or physical, whether of occult qualities or mechanical, have no place in experimental philosophy."

There was considerable reaction to this point of view. When the second edition of *Principia* was prepared beginning in 1708, Roger Cotes, who superintended its preparation and wrote a preface for the edition, begged Newton to change the section about hypotheses. He wrote about "the last sheet of your Book which is not yet printed off"; he wanted it revised because he could not "undertake to answer anyone who should assert that you do *Hypothesim fingero . . .*" Newton never made the requested changes, but in his *Opticks* Newton presents plenty of hypotheses not only on the subject of gravitation, but also on many others. This inconsistency has given scholars and philosophers fuel with which to write in detail about Newton's meaning of the term *hypothesis* when it could well be nothing more than a mistake of dogmatism, maintaining his position in spite of logical opposition to his view.

Newton's dogmatic attitude may not have served the best interests of science, and neither did the uncritical attitude of his followers. The process of almost apotheosizing Newton began in his time, and whatever he said was taken as gospel. For example, he said that glass of any kind had the same refractive index, and this was accepted as absolute truth. No attempt was made by opticians to combine lenses of different kinds to change the refractive index.

The dogma of Newtonianism grew as Newton was applauded and honored. Alexander Pope wrote his famous couplet about nature and her laws being "hid in night until God created Newton and all was light." Christian Huyghens, who died in 1695, helped this along with his contention, put to use by the followers of Newton, that "in true philosophy, one conceives the cause of all natural effects in terms of mechanical motions."

Eighteenth- and nineteenth-century scientists saw the physical world in terms of Newtonian principles. The Marquis de la Place, at the end of the eighteenth century, espoused the extreme faith in Newtonianism with his statement that all could be solved given sufficient data. He said: "An intelligence who knew at a given moment all the forces existing in nature and the relative position of all existing things or elements composing it, would, if he were able to submit all these data to mathematical analysis, be able to comprehend in a single formula the motion of the greatest heavenly body and of the lightest atom; nothing would be uncertain for him, and future as well as past would be open before his eyes." In the nineteenth century, German physicist Hermann von Helmholtz claimed that the final goal of all science was to become a branch of Newtonian mechanics. This was echoed by his compatriot Kirchhoff, who said: "The highest object at which the natural sciences are constrained to aim . . . is the reduction of all the phenomena of nature to mechanics." The trend continued into the twentieth century when French mathematician Paul Painlevé said, "mechanics is the necessary foundation for the other sciences." Yet today, Newtonian mechanics is no longer the basis for quantum, relativity, and field theories.

Dogmatism still pervades science. The illustrious twentieth-century philosopher and mathematician Alfred North Whitehead maintained that scientists are what large segments of the clergy were a few generations earlier—"the standing examples of obscurantism." In 1939, he called the scientists the chief representatives of that "self-satisfied dogmatism with which mankind at each period of its history cherishes the delusion of the finality of its existing modes of knowledge." He found "after escaping the certainty and dogma of ecclesiastics [that] the scientists"—from whom he had expected an elastic and liberal outlook—"were the same people in a different setting." Another philosopher, M. Bunge of Freiburg University, dissatisfied with quantum theory interpretations, said: "For the first time in history, scientists have managed to outdogmatize philosophers."

Dogmatism in science is often a very common and not especially sophisticated type. The Velikovsky affair, described in Chapter Five, was brought on by dogmatists. In September 1969, following an American Chemical Society symposium on "Public Policy Aspects of Environmental Improvement," a member of the society told the editor of *Chemical and Engineer-*

ing News that one of the speakers had no business being invited to an American Chemical Society platform.

Appropriate education can reduce the dogmatism in science. A scientific education can no longer be one where training in a scientific discipline is assumed sufficient for the growth of the individual. If future scientists are to be open-minded and tolerant, they need to be aware of the history and interactions of science. Courses in the history of science as such, more so than the history of chemistry or of psychology, can reveal the morals to be absorbed by the future scientist.

SELECTED REFERENCES

Cohen, I. B. "Newton in the Light of Recent Scholarship," *Isis,* 51 (December, 1960), 489–514.

Goran, M. "The Promise of Scientific Humanism," *School Science and Mathematics,* 70 (October, 1970), 629–634.

———. "Towards Scientific Humanism," *Journal of Higher Education,* 14 (Autumn, 1943), 435–438.

More, L. T. *Isaac Newton.* New York: Scribners, 1934.

Newton, I. *Opticks.* New York: Dover, 1952.

O'Malley, C. D. *Andreas Vesalius of Brussels.* Berkeley: University of California Press, 1964.

9

ACCENT THE POSITIVE

Laughter generally ensues when one reads of cases where non-scientists have made negative predictions about science and technology which later prove dead wrong. One week before the success of the Wright brothers at Kitty Hawk, a *New York Times* editorial advised: "We hope that Professor Langley will not put his substantial greatness as a scientist in further peril by continuing to waste his time, and the money involved, in further airship experiments. Life is short, and he is capable of services to humanity incomparably greater than can be expected to result from trying to fly." Admiral William Leahy said in 1945: "That is the biggest fool thing we have ever done. The [atomic] bomb will never go off, and I speak as an expert in explosives." When the telephone was first invented, the chief engineer of the British Post Office was asked about it, to which he replied: "The Americans have need of the telephone, but we do not—we have plenty of messenger boys."

Scientists are not expected to predict wrongly, but sometimes they do. At the start of the twentieth century, the father of nuclear science, Ernest, Lord Rutherford called discussions of the scientific method and of the philosophy of science "all hot air." New Zealander Rutherford was an exceptionally brilliant experimental physicist who established the laws of radioactive decay, formulated the general structure of the atomic nucleus, and fostered scientific research in England. Toward the end of his long and useful career, he claimed that man would never tap the energy of the nucleus; the atomic bomb was exploded a few years after his death.

Rutherford has a great deal of company in the class of scientists who make negative dogmatic statements. There are others whose pessimism about man's ability is recorded, and it was often particularly unfortunate in its timing. A leading physiologist in Germany in 1846, Johannes Peter Müller, said that the speed of nerve conduction would never be measured; six years later, Hermann von Helmholtz measured the speed of nerve impulse in a segment of frog nerve. When electricity was being developed in the United States, Professor Elihu Thompson once told newspaper reporters that he, an expert in electricity, "did not think very highly of the Edison lamp and expected no great future for it." When Max Planck began his career in physics in 1875, he was told by the head of the physics department of the University of Munich (where his father taught law) that "physics is a branch of knowledge that is just about complete. The important discoveries, all of them, have been made. It is hardly worth entering physics anymore." The American experimental physicist who accurately measured the speed of light, Albert Michelson, was another who claimed that the important ideas of physics were known and the task of a physicist was simply to make more accurate measurements. Soon thereafter, physics was revolutionized by relativity and quantum theories as well as by the discoveries of x-rays and radioactivity and by the isolation of the electron.

The nineteenth-century Scotch physicist Kelvin scoffed at the idea of atomic structure, claiming that the word *atom* is by definition a thing that cannot be divided. One apocryphal story has a student questioning him on this matter say, "That shows the disadvantage of knowing Greek." Kelvin, however, was not alone in being dogmatically opposed to atomic theory. Humphry Davy was openly opposed, although he later accepted the thesis. Likewise unconvinced was one-time Harvard University president Charles W. Eliot. When teaching chemistry, he would tell his students that "the existence of atoms is itself a hypothesis, and not a probable one." Ernst Mach, an Austrian whose *Science of Mechanics* inspired Albert Einstein as well as the philosophical school of positivism, said, "In my old age, I can accept the theory of relativity just as little as I can accept the existence of atoms and other such dogma."

Shortly before the Wright brothers successfully demonstrated a plane in flight at Kitty Hawk, North Carolina, American astronomer Simon Newcomb said that man would never fly. An article in the November 1909 *American Review of Reviews* argued confidently that the airplane was a toy. George Eastman, the photography pioneer, said about the talking picture: "I wouldn't give a dime for all the possibilities of that invention. The public will never accept it."

During the 1920's, a German science student at Columbia University heard from one of his eminent teachers that the student's idea of an elementary particle of matter, such as an electron, spinning was impossible. Before Ernest O. Lawrence of the University of California built his particle-accelerating device, the cyclotron, a good friend and colleague, later an important nuclear physicist, had told him that it was impossible.

In 1936, distinguished Austrian physicist Hans Thirring contended that there could be no space flight. In 1945, despite the evidence of the German V-2 rockets, Vannevar Bush, a dean of American science, told a U.S. Senate Committee: "The people who have been writing these things that annoy me, have been talking about a 3,000-mile high angle rocket shot from one continent to another, carrying an atomic bomb and so directed as to be a precise weapon which would land exactly on target . . . I say, technically, I don't think anyone in the world knows how to do such a thing, and I feel confident that it will not be done for a very long time to come." In 1967, Sir John Eccles, the Nobel-prize-winning brain physiologist, claimed that extraterrestrial intelligence did not exist. Travel between the stars, he said, is out for all time.

Dogmatic assertions about the impossibility of one thing or another can limit scientific research when such seemingly authoritative predictions as those cited above are taken too seriously. Young scientists do sometimes idolize their teachers, and they may pick up some of the teachers' negative approaches, thus closing themselves off from positive views.

Some currently accepted concepts are often stated in negative terms, and they therefore impose a similar barrier to the acquisition of new knowledge. The laws of conservation of matter and energy, for example, are glibly recited as "matter cannot be created nor destroyed" and "energy cannot be created nor destroyed." Aside from the inability to define *creation* and *destruction* in scientific terms, the use of the negative approach is not conducive to best understanding or to research. The same idea can be stated more effectively in positive terms: "The amount of energy or matter in a closed system is constant, but transformations may occur."

Some consequences of scientific principles are also often given in negative terms. Archimedes principle, for example, is that objects immersed in a fluid are pushed up by the weight of the fluid displaced. A stone in water is buoyed up by the weight of the water taken up by the volume of the stone, or a balloon is lifted by the weight of air defined by the volume of the balloon. Some say that the principle indicates that balloons cannot be used to explore outer space because the balloon's weight becomes equal to the buoyant force not far from the surface of the earth. However, it may be that some day rockets will shoot uninflated balloons into outer space, where chemicals will inflate them, as was the case of the Echo satellites, and the statement that balloons cannot explore outer space will become subject to qualification. Were the original statement more positive, the difficulty would disappear.

Facts of science can be stated in a negative fashion, but the positive view is more fruitful and complete. It is possible to say without qualification that "The entire earth is not a flat surface," but understanding is enhanced if the negative is avoided and the statment is that "The earth is shaped like an oblate spheroid." Again, the greatest density of water can be described with the statement "It is not at the freezing point," which tells only a part of the story; whereas the statement "The greatest density of water is at 4° centigrade," tells the whole story.

The most damaging kind of negative approach is toward the scientific enterprise. This negative attitude toward science indicates an inability to solve particular problems or entire sets of problems. But the scientist unwittingly assumes that he and others are capable of solving problems; theirs is an optimistic view of the powers of mankind. Whatever difficulties are present they regard as temporary obstacles which will be overcome eventually. The individual scientist may temporarily or permanently abandon a tough problem; yet others of his and succeeding generations will probably tackle it and make a contribution. Rudolf Virchow, the father of modern pathology and a German contemporary of Darwin, proclaimed "knowledge has no boundary other than ignorance." This is a point of view contrary to that held by those who consider nature inherently indeterminate, unpredictable, and unknowable.

Hardly anyone in science today is so optimistic that he envisions the time when all problems will be solved. Such a goal of a closed system where all the answers are known and problems are nonexistent is the substitution of an unscientific or antiscientific setting for the dynamic, open system character of science.

The idea that progress is at an end seems to reappear many times in the history of civilization. The first U.S. Commissioner of Patents, Henry L. Ellsworth, wrote in 1843, when new patents totaled less than five hundred a year: "The advancement of the arts from year to year taxes our credulity and seems to presage the arrival of that period when human improvement must end." (In August 1966, the U.S. Patent Office issued 1,400 patents in one week.) Later in the nineteenth century, some physicists spread the doctrine that their work henceforth would be that of expert measure. James Clerk Maxwell clearly saw the meaning of this. He said that if the condition were approaching, then the science laboratory would have to be just another "great workshop of our country, where equal ability is directed to more useful ends. But we have no right to think thus of the unsearchable riches of creation." He said that the history of science could not support the view that discovery was at an end.

In our time, environmental and social problems have made negative and pessimistic forecasts more fashionable. Scientists of reputation join in a chorus prophesying doom not only for mankind, but also for living kind. Some advocate a moratorium on science and invention. Others counter with the opposite extreme of scientism, claiming that science will solve all our problems.

Effective communication and appropriate education in the sciences and other disciplines will not solve all our problems, but they will show the dead end the negative approach leads to as well as the hope that is engendered by viewing problems with confidence in a possible solution.

SELECTED REFERENCES

Barzun, J. *Science: The Glorious Entertainment*. New York: Harper & Row, 1964.

Clarke, A. C. *Profiles of the Future: An Inquiry into the Limits of the Possible*. New York: Harper & Row, 1963.

Kahn, H., and Weiner, A. J. *The Year 2000*. New York: Macmillan, 1967.

10
IGNORING GENIUS

Citing the successes who were failures at school may be a device to bolster the confidence of the average and poor student or to knock organized education. The practice is widespread and an amazing array of evidence has accumulated. The number of illiterates who became financially successful is impressive, although this phenomenon is more characteristic of earlier times in American life. The self-taught as well as the expert who became so, contrary to the expectations of his teachers, are in every profession, craft, and business activity; science offers no exception. It is now commonplace to remark that Einstein flunked a mathematics course; he was also the equivalent of a modern high school dropout. One of his high school teachers told him that he would never amount to anything. Then too, he failed the entrance examination to the Zurich Polytechnical Institute. At the beginning of the nineteenth century, brilliant mathematician Evariste Galois twice failed the entrance examination given by the École Polytechnic. Later in the century, the founder of modern genetics, Mendel, was a poor student; Darwin too was a dullard in school. The developer of radio, Guglielmo Marconi, was so incompetent a student that he was unable to matriculate at the University of Bologna.

The teachers of great scientists often have not recognized their genius, even when their school work was not failing. The father of Enrico Fermi believed that the schools knew his son as a good student but not as the prodigy he was. Perhaps the accent of schools on the acquisition of knowledge rather than on the development of critical and creative thought has prevented such recognition. A contemporary nuclear scientist has confessed that his mother succeeded in instilling practice in thought in him not by asking at the end of the elementary school day the classical "What did you learn today?" but rather by inquiring "What questions did you ask today?" Formulating proper questions is considered by many to be a key procedure of science.

Organized education is, however, responsible for moving a large number of great scientists into scientific work and for adequately preparing many more average scientists. The record is clear: Inspiring teachers and books have done much for science. Michael Faraday became eminent in nineteenth-century physics without having a university training, but he had the good fortune of having Humphry Davy as an employer and tutor. Faraday said that early in life he had read Isaac Watts's *The Improvement of the Mind,* and he singled it out as the starting point of his intellectual development. Watts's book advised such activities as attending lectures, corresponding, keeping a notebook of interesting facts and ideas, and participating in small discussion groups. Faraday also gave credit to Mrs. Marat as the author of *Conversations on Chemistry* as "one who had conferred great personal good and pleasure on me; and then as one able to convey the truth and principle of those boundless fields of knowledge which concern natural things to the young, untaught, and inquiring mind." A contemporary dean of American chemistry, Joel Hildebrand of the University of California, gave some credit for starting him in the scientific way of life to his high school principal, who had entrusted him with a key to the laboratory. Glenn Seaborg, chairman of the Atomic Energy Commission during the 1960's, said his high school chemistry teacher had inspired him to become a scientist. Evidently the problems of science education would be eased considerably if inspiring teachers and books became more abundant.

No matter what mechanism may have started people off in scientific work, it is clear that not all of them, even those of superior accomplishment, have always been recognized by the scientific community. A classic case is that of Edward Jenner; he submitted a report on cowpox vaccination to the Royal Society and it was rejected with the admonition that "he should not risk his good reputation by publishing such an experiment." In 1802, however, Parliament gave Jenner 10,000 pounds, and five years later he received an additional 20,000 pounds.

The reasons for the failure to recognize genius are many and varied in any given situation. Sometimes it is incompetence on the part of the judges. In other instances the reason may be inertia and resistance to change. Perhaps fear of Church or state authorities may be a factor. Perhaps that was the reason for René Descartes' failure to visit Galileo; Descartes said he saw nothing in Galileo's writings he would care to call his own. Some cases can be assigned to economic conditions. Only lately has there been a great market for scientific talent; prior to the twentieth century, science had no organized system of support and it was difficult for a scientist to find a job. Between 1875 and 1878, David Starr Jordan, an expert on fishes who became president of Stanford University (and president of the American Association for the Advancement of Science in 1909), had tried early in his career to get a job at nine different institutions and failed. He taught at an Indianapolis high school and at Northwestern Christian University—later Butler University. The director of the Peabody Academy of Science at Salem, Massachusetts, did not care to recommend him for an appointment at the University of Cincinnati; the latter turned him down with the statement that they already had three people from Cornell University, his alma mater, among their dozen or so faculty members.

Failure to recognize genius can be due to inadvertence. In 1829, Evariste Galois, seventeen years old, sent his first communication to the French Academy, and it was lost through the negligence of Cauchy, the secretary. Between the time Galois was eighteen and the time when he died at twenty, he sent more papers to the Academy. One was taken home by Fourier, the permanent secretary, but he died before examining the contents. The same fate of having an important communication overlooked befell Robert Mayer, an independent co-discoverer of the law of conservation of energy; he sent a report to Poggendorf's *Annalen der Physik,* but it was found unopened years later at Poggendorf's death.

The mistake of not noticing superior accomplishment can be due to the sophistry of believing that only centers of excellence can bring forth products worthy of attention and study. It occurs today when scientists willingly give their attention to anything that comes from large and impressive universities but must be pushed into reading a contribution that comes from a lesser-known school. The so-called first-rate journals carry the work done at the top universities and rarely, if ever, have reports from Podunk College. At scientific meetings too, the normal association of scientists enhances the invisible "establishment." Accordingly, when someone outside the "circle" does have talent in research, it takes a longer time to recognize. The young scientist at a so-called second-rate school may become cynical and adopt the slogan that "It's who you know rather than what you know" that brings recognition. A case in point occurred in the middle of the nineteenth century when J. J. Waterson, a resident of Bombay, India, sent what is essentially a version of the kinetic molecular theory to the Royal Society. The paper went into the archives of the organization, where it was rescued in 1892. Waterson had brought his ideas before the British Association for the Advancement of Science in 1851 and had described them in an article in the *Philosophical Magazine* in 1858, but they had not been noticed. In 1928, an edition of Waterson's work was issued and the editor, J. S. Haldane, wrote: "It is possible that, in the long and honorable history of the Royal Society, no mistake more disastrous in its actual consequences for the progress of science and the reputation of science than the rejection of Waterson's paper was ever made." The referees for publication had commented a century ago "the paper is nothing but nonsense."

Inability to recognize scientific talent is a fault that may be rationalized away with the claim that society was not ready for the genius and his ideas; he was far ahead of his time. However, it is also true that upholders of the status quo are not receptive to revolutionary thoughts. The innovator rocks the boat, and those who control the vessel do not solicit any upsetting forces. In the nineteenth century, Hermann von Helmholtz wrote that "the greatest benefactors of mankind usually do not obtain a full reward during their lifetime," and the "new ideas need more time for gaining general assent the more really original they are." Shortly thereafter, the sea of science was host to a large number of boat rockers who received little or no recognition during their lifetime. Nicholas Léonard Sadi Carnot is a good example.

In 1824, Carnot published a 128-page book in Paris, *Reflections on the Motive Power of Heat and on the Machines Adapted to Develop This Power*. This volume established the science of thermodynamics and prompted a late nineteenth-century physical scientist to suggest that Carnot was the most brilliant physical scientist of the time. The failure of scientists to recognize Carnot's genius until many years after he was dead is all the more remarkable in light of his family connections. Carnot was not a member of the lower classes seeking fame and fortune; he was from an established upper-class family. He was born in 1796 in the Luxembourg Palace in Paris. His father had once served as Napoleon's Minister of War, and at the time of Sadi's birth, he was a member of the Directory. Sadi's younger brother also had a son called Sadi who became the fourth president of the Third Republic. Sadi Carnot, the future scientist, graduated from the École Polytechnic and received a commission in the engineer corps. He was appointed to the general staff in Paris, where he had plenty of leisure for study and research. He resigned from the army eventually but continued his scientific work until he died from cholera at the age of thirty-six.

At the beginning of his important book, Carnot described the immediate significance of his work. He wrote: "Already the steam engine works our mines, impels our ships, excavates our ports and our rivers, forges iron, fashions wood, etc. . . . It appears that it must some day serve as a universal meter, and be substituted for animal power, waterfalls, and air currents . . . Yet, its theory is very little understood, and the attempts to improve it are still directed almost by chance." He wrote as an engineer, but he was neglected by engineers as well as by the scientists. As a result of his study of heat engines, Carnot advised the greatest possible temperature difference between boiler and condenser. It was not the use of steam, alcohol, or any other medium that so vitally affected the efficiency of the steam engine; he found that the efficiency increased when the difference in temperature between intake and exhaust increased.

In 1834, a classmate of Sadi Carnot, the French civil engineer

Clapeyron translated Carnot's ideas into a more mathematical form. But his paper, as well as Carnot's book, was ignored by practicing engineers. In 1852, the celebrated engineer John Ericsson tried, at great cost, to build a large so-called atmospheric engine, supposed to be able to operate without a temperature difference between intake and exhaust.

Carnot was rescued from obscurity, independently, by the physicists William Thomson and Rudolf Clausius. In 1845, twenty-one-year-old Thomson, later to become Lord Kelvin, was working in Regnault's laboratory in Paris; he was interested in getting more accurate thermal data on steam engines. Thomson described his adventures trying to locate a copy of Carnot's book: "I went to every bookshop I could think of, asking for the *Puissance Matrice du feu* by Carnot. They produced a volume on some social question by Hippolyte Carnot but the *Puissance* was quite unknown."

When Hippolyte Carnot, Sadi's younger brother, brought the original manuscript of the book to the French Academy in 1878, he also presented them with parts of Carnot's notebook. These fragments indicate that Carnot was thinking about and could have established the law of conservation of energy. Carnot calculated the mechanical equivalent of heat (and came close to the accepted value). Some scholars credit him with the first statement of the second law of thermodynamics, the statement that unavailable energy in any closed system is coming to a maximum.

Another genius who stated the first law of thermodynamics, the law of conservation of energy, was also ignored by his contemporaries. In 1840, a twenty-six-year-old physician, Robert Mayer, was on a ship sailing to the Dutch East Indies region; it was then that he arrived at a statement of the idea. His description was published in 1842 in Liebig's *Annalen der Chemie und Pharmacie,* but it received hardly any attention. Mayer became a physician in his native city of Heilbronn, where some of his fellow citizens ridiculed his claims to scientific creativity. Perhaps this was a reason for his spending some time in an institution for the insane.

A third scientist who formulated the conservation of energy law was Hermann von Helmholtz. In 1847, his paper "On the

Conservation of Energy" was refused for publication by the leading journal of physics. It had to be presented to the world of learning as a pamphlet.

Sometimes the failure to recognize genius is caused by professional jealousy. Perhaps this is why Francis Bacon denounced William Gilbert's *On the Magnet* as "a work of inconclusive writing" and full of "fables." Jealousy is more evident in the case of nineteenth-century Hungarian physician Ignatz Semmelweis. Upon receiving his medical degree from the University of Vienna in 1844, he received an appointment in the maternity department of a hospital directed by Dr. Johann Klein. At that time, the exceptionally high death rate in the maternity wards was assigned to fear, atmospheric agents, overcrowding, and other such factors. Semmelweis came to believe it was caused by improper sanitation practices on the part of visiting students.

In 1847, he suggested that chlorinated limewater be used by students when washing their hands. This was adopted, and by the end of the year, the mortality rate in his department fell to three and a fraction percent from a high of twelve and one-quarter percent; within another year the percentage fell to one and one-quarter percent. Semmelweis did get some support as a result of this achievement, but the jealousy of Klein came to the fore. This along with Semmelweis' lack of tact caused him to be discharged. He did find another post in the maternity department of another hospital, and again the antiseptic measures he suggested dramatically reduced the mortality rate. Semmelweis died in an insane asylum after six years of successful work in this hospital.

Another similar case may be that of the English scientist J. R. Newlands. He produced an idea that was a forerunner of the periodic table of chemical elements credited to the Russian Dmitri Mendeleef. Newlands proposed a law of octaves, the arrangement of elements in groups of eight to show recurrence of properties. His contemporaries sarcastically asked why he did not arrange the elements in alphabetical order, and he was veritably laughed out of scientific work. Another comparable case may be that of the Russian botanist Tswett, who in 1906 presented the basis for chromatography, now widely used for chemical analysis. The idea did not take hold then, and Tswett's life was not only one of poverty, but also of nonrecognition.

It is not always possible to assign causes to complex issues involving personalities. Perhaps a measure of professional jealousy was involved in all situations where superb work was not recognized. Documentation of such a charge is difficult. Motives may not be overtly expressed, and the written expression cannot be trusted to be real.

Sometimes a scientist and his accomplishments are overshadowed by a strong personality with excellent press relations. That may have been the case for Thomas Hariot (1560–1621), who may have used a telescope before Galileo did. Hariot was a physicist interested in falling bodies, a mathematician, and an astronomical observer who made the earliest known observations of Jupiter's satellites. But he lived under the shadow of Galileo's prowess as Robert Hooke lived under the shadow of Newton's great talents.

The story of the monk Gregor Mendel is representative of the cases where the full reasons for neglect have not been clearly established. Mendel was born in Silesia, then a part of Austria, in 1822. His parents were so poor that his youngest sister had to give him her dowry in order to finance his education. He became a secondary school teacher, although he twice failed the examination for teachers. He did make an impression on one of his examiners who persuaded the Abbott to send Mendel to the University of Vienna. Following this training he became a teacher at a Brünn school, and this is where he did his breeding experiments with garden peas.

Until Mendel did his work, there were only empirical rules about heredity. The breeder of cattle and the farmer were responsible for such observations as "The apple doesn't fall far from the tree" and "As the twig is bent, so grows the tree." There was also a large body of opinion that assigned the hereditary process to the blood; this idea still lingers in common parlance with the expression "It's in his blood." Mendel selected to observe three easily distinguishable characteristics of garden peas: color, size, and quality of coat. Beginning with seeds that were pure and never mixed, Mendel mated them and noted whether the offspring were tall or short, yellow or green, with smooth coat or wrinkled coat.

Mendel's work was rediscovered at the beginning of the twentieth century by three different scientists independently. One of them, Carl Correns, was rebuffed by a botanist at Tübingen University who had discovered osmosis, the flow of dissolving material from more to less concentrated regions of it. This botanist had tried to discuss osmosis with a professor of physics at his university, but the physicist would not listen. It would seem that the botanist, Wilhelm Pfeffer, would thus be more tolerant to other new ideas. But when Correns asked Pfeffer for laboratory space to study heredity, Pfeffer turned him down with the remark that horticulture was not worthy of a botanist.

There are a number of possible reasons why Mendel's work was not acknowledged during his lifetime, but the one usually given—that the journal in which his results were published was too obscure—is just not true. The magazine was well circulated for the time. The usual inadvertent neglect may be one of the true causes. Mendel sent a reprint of his article to the Viennese botanist Kerner, who did not even cut the pages in order to read the report. Mendel corresponded with Karl Nägeli, a distinguished professor of botany at Tübingen University, who replied in a condescending tone. Although Mendel had used more than ten thousand plants, Nägeli advised that the experiments were not complete. However, he did indicate a willingness to repeat the experiments if he were provided with the seed.

Perhaps a major cause of Mendel's neglect is that biologists specifically and scientists in general were not yet accustomed to a concept as imponderable as the gene, the unit of heredity. At that time the atomic idea was being resisted by some well-known scientists. If the physical scientists had such problems, then the biologists had even more acute ones. The physical scientists were more attuned to accepting intangibles. They had learned about phlogiston, the intangible supposedly a part of all burning, and caloric, the intangible essence of heat. On the other hand, biologists simply described the perceivable natural world. Even when such instruments as the microscope became available, biologists simply observed more carefully. Application of the concepts of physics and chemistry was rare. An added comparable difficulty was the statistical and quantitative nature of Mendel's results. Biology, until Mendel was rediscovered, had no use for numbers and mathematical analyses. Truly, some forward-looking scientists did measure the heat generated by animals, but these scientists were considered chemists and physicists, not biologists.

Perhaps the contemporaries of Mendel were prejudiced against the clergy, or at least they did not expect major scientific advances from them. Prior to Galileo, the Church had been the economic support of those who had fostered some science: Roger Bacon, Robert Grosseteste, Albertus Magnus, Nicholas Copernicus, and others. During Galileo's day, some churchmen had been active in research. Father Clavius, for example, was an astronomer who had a sizeable moon crater named after him. But as soon as the scientific revolution proceeded after Newton—religious but not a member of the Church hierarchy—the Church was in a different position. Scientific investigators began to receive support from universities and wealthy patrons. For a century and a half after Newton's death, scientists were not accustomed to having a distinctive scientific advance emanate from a churchman. Perhaps they were not attuned in this respect to receive Mendel.

Perhaps the scientists were prejudiced against a Silesian. The records of history reveal that scientific and technical advance is the product of varying kinds of people. No one country has a monopoly on scientific talent. However, scientists are products of their societies and may be prejudiced. (A favorite expression of Thomas Jefferson was that "the failure of almost all the great scientific or literary undertakings of Americans is to be attributed to their employment of foreigners, instead of calling into exercise the talents of their own citizens.") It could be that the dominant circles of science in Mendel's day, French, German, and English, were prejudiced against any other nationality—especially those without great names in the history of science. Silesia was close to eastern Europe, where the tenor of civilization was not at all like that of the West.

In Mendel's time the scientific rivalry between Justus Liebig, a German, and Louis Pasteur, a Frenchman, was sometimes spiced with overtones of nationalism. When told about Pasteur's theory that microbes could cause fermentation, Liebig dismissed it as naïve and childish and comparable to the theory of one "who would explain the rapidity of the Rhine current by attributing it to the violent movement of the many mill wheels at Mainz." Pasteur wrote about some of his experiments on fermentation: "I undertook them immediately after the war of 1870, and have since continued them without interruption, with the determination of perfecting them, and thereby benefiting a branch of industry (brewing) where we are undoubtedly surpassed by Germany." During the twentieth century, Lord Rutherford would say, perhaps jokingly, that no Anglo-Saxon could understand relativity. In 1965, French Nobel-prize winner in medicine Jacques Monod said, "Scientific chauvinism, so strong in France and, by the way, in Britain and Germany, does not exist in the United States." In 1966, Belgium's age-old dispute over whether Flemish or French should be the national language seriously affected the country's scientific research. The University of Louvain, said to be the largest Catholic university in the world, was split into two faculties, French-speaking and Flemish-speaking, and some research duplication resulted.

Another factor that worked against Mendel's recognition while he lived was timing. It was the age when evolution and Darwinism were being discussed. There seemed at the time to be no connection between Mendel's work and evolution.

Will genius in science be ignored at times in the future? The answer is obviously yes because the factors that existed in Mendel's time are still in operation in pure and applied science. Max Planck in his *Scientific Autobiography* relates how late in the nineteenth century he had difficulty in obtaining an appreciation of his work. Helmholtz probably did not read the papers, and Kirchhoff disapproved; Clausius did not answer Planck's letters and was not at home when Planck called in person. He corresponded with Carl Neumann of Leipzig but "it remained totally fruitless."

Consider the case of Julius E. Lilienfeld, a native of Poland and formerly a professor of physics at the University of Leipzig;

he became a citizen of the United States in 1935. Lilienfeld was granted U.S. patents in 1930, 1932, and 1935 for a kind of solid-state amplifier, more than two decades before it was rediscovered. Lilienfeld died in 1963, when the modern transistor was being used and appreciated; his work with amplifying devices having the basic characteristics of the modern transistor came to be noted only after his death. Another recent case involves the Van Allen radiation belt, named for the University of Iowa physicist James Van Allen although its existence was noted in 1935 by E. M. Bruins of the Netherlands in his doctoral dissertation, which had a long summary in English; four hundred copies had been sent all over the world.

There are modern cases with happier endings. Chester Carlson, the inventor of a dry-copy machine, was turned down by twenty large organizations before a small one agreed to try his device; they marketed it as Xerox. In 1952, a group at Brookhaven National Laboratory invented a principle of magnetic focusing for particle accelerators. But their method had been independently invented by an American-born Greek, N. Christofilos, who was employed in Greece selling elevators for an American company; he was a physicist in his spare time. Christofilos had sent a manuscript describing his invention to the national laboratory at Berkeley, California, where it was forgotten. Then news of the work at Brookhaven reminded someone at Berkeley of the Christofilos manuscript. Christofilos thereupon secured employment with the Livermore Laboratory at the University of California.

Because its causes are many and difficult to control, failure to recognize genius in science will happen in the future. However, the occurrence can be decreased through international funding of science. The channeling of money to a variety of countries with qualified investigators ought to diminish the number of scientific workers outside the mainstream of the scientific community. Effective communication can also help alleviate the problem. When abstracts of results, no matter the journal of communication, are available to all, the chances for an oversight become smaller. Education may be the most important factor diminishing the failure to recognize genius. Science students, imbued with a history of their subject and a firm commitment to it, will be motivated to avoid ignoring any accomplishment.

SELECTED REFERENCES

Barber, B. "Resistance by Scientists to Scientific Discovery," *Science,* 134 (September, 1961), 596–602.

Goran, M. "Public Relations and the Heroes of Science," *The Chemist,* 44 (November, 1967), 397–98.

Gordon, T. J. *Ideas in Conflict.* New York: St. Martin's Press, 1966.

Kerker, M. "Sadi Carnot and the Steam Engine Engineers," *Isis,* 51 (September, 1960), 257–70.

Koenig, F. O. "On the History of Science and of the Second Law of Thermodynamics," in H. Evans, Ed., *Men and Moments in the History of Science.* Seattle: University of Washington Press, 1959.

11
RECOGNITION

Early in the twentieth century, the American sociologist William Isaac Thomas published his concept of the four wishes desired by all men and women. He called them the wishes for new experience, security, response, and recognition. (He defined response as the need for love and affection.) New analysts of human behavior suggest that all individuals seek some form of security, new experiences, expression, commitment, recognition, and perhaps some accomplishment; new analysts consider the drive for love and affection either major and separate, or part of the others. These, in one mixture or another, in one form or another, satisfy human needs. The security may be economic, emotional, and even political; the new experiences can simply be escape from routine and boredom; the expression makes possible the communication of feelings and ideas; commitment gives values to live by and for; the recognition is that which establishes esteem, either self-esteem or the respect of others. Shakespeare's "good name in man or woman is the immediate jewel of their soul; that away, and they are but gilded loam or painted clay" is as true in science as in any other endeavor.

Recognition for a scientist is often the recognition by his peers that he has made and continues to make a contribution to science. This seemingly simple matter has led to outrageous behavior by some scientists. Their abnormal need for acclaim, or an innate inability to know when it is given, has resulted in some very unpleasant incidents in the history of science. The scientist who is aware of the cooperative nature of scientific discovery should, theoretically at least, be more sensible about taking and acknowledging credit. Those alert to social patterns can be tolerant and agree with the Nobel-prize winner in physics who in 1966 said: "The world is peculiar in this matter of how it gives credit. It tends to give credit to already famous people." But the mistake of being very greedy for acclaim has been found in the great, near great, and ordinary scientists since the beginning of modern science.

The roots of the scientific procedures of the present era can be traced back earlier than man's existence. Some procedures of science have their counterparts in the animal world. Such activity as hypothesis formation and verification, tool using and building, and effective use of imagination can be assigned to primates and to less-specialized organisms. The development of our present prowess can be viewed as an accumulation of techniques and patterns since early animal times. The theory of organic evolution is in this perspective a record of successful problem-solving by species that developed into man. So if due credit is to be assigned, man's nervous system and cerebral cortex, his central agency for problem-solving, arose initially out of the experience of lower organisms.

It is more customary to begin the story of the rise of man by citing the achievements of early man. These were considerable, from the invention of the wheel to the domestication of animals. No one individual is given credit for these important advances; they are assigned to the group. As the record of civilization becomes more specific and definite, individual scholars and scientists are awarded high honors for their successes. The names of Archimedes, Aristotle, and Galen, for example, are representative of early Greek and Roman achievers. There is no problem of assigning credit. A similar situation of not having individuals to whom to assign credit occurred in early Christian times, the Middle Ages, among the Arabs in the Middle Ages, and even in the period preceding the Renaissance. It may be that the Renaissance so elevated the individual that personal achievement became important, and the problem of who should be credited with what then became a serious matter.

At the beginning of modern science, in Galileo's time, the abnormal drive for credit affected many of the best scientists.

René Descartes, the French founder of analytical geometry, the philosopher who built a system based upon the certainty of man's thought, the physicist whose ideas were supplanted by Newton's, failed to credit one of his best sources for ideas. Isaac Beeckman of Holland, first a student of theology, then a candle maker, pipe layer in breweries, student of medicine and the arts, and finally Rector of Latin Schools in Holland, was for some time a friend and confidante of Descartes. Beeckman's diary was discovered in 1905, and it reveals the indebtedness of Descartes. When leaving Beeckman for the last time, Descartes wrote: "You are truly the only one who has awakened a lazy man, recalled a learning that had already almost escaped his memory, and brought back to a better road a mind turned away from serious things. And if by chance I come one day to produce something that was not contemptible, you have the right to claim the whole of it; and I shall not omit to send it to you, that you may enjoy or correct it." Beeckman died of consumption when not yet fifty, on May 10, 1637; Descartes' first essays were then published. Descartes denied any debt to Beeckman; he so wrote Father Marin Mersenne, the scientist-clergyman who served as the equivalent of a modern scientific journal. Experimenters sent letters to him and even made him a confidante. (Mersenne's own scientific accomplishments were in the study of sound and the mathematics of music.)

A more striking lack of assigning due credit is contained in the Galileo and Kepler relationship. Their first contact was during the spring of 1597, when Kepler sent Galileo a copy of *Mystery of the Universe,* Kepler's first book. During the summer of the same year, Galileo answered. He had read only the preface, he wrote, but felt a great joy in meeting so powerful an associate in the pursuit of truth. He concluded his letter by confessing that he had not yet dared to publish his findings, fearing the fate of Copernicus. He wished there were more like Kepler, for then, he, Galileo, would venture to publish his speculations. In Kepler's answer, dated October 13, 1597, Galileo was invited to publish in Germany. Kepler asked Galileo for several observations because Kepler did not have the instrument necessary, an angle-measuring one called a quadrant. There is no record of Galileo ever having supplied this information. The correspondence between the two seems to have ended abruptly. Thirteen years later however, a friend of Kepler's reported to him that Galileo had used a telescope to make some new discoveries. A month after this, Kepler received a copy of Galileo's *Starry Messenger* through the Tuscan ambassador in Prague. Kepler wrote Galileo: "I long for a telescope." In August of the same year, Kepler wrote again, expressing his desire to see Galileo's instrument so that he too might enjoy "the great performance in the sky." Ten days later, on August 19, 1610, Galileo answered Kepler's letter, but it was not a full reply. In it Galileo mocked the philosophers for their stubbornness in refusing to see the planets and moons through his telescope. Galileo thanked Kepler for having been the first and almost the only one with frankness and intellectual superiority to have given his full support without having seen the objects through the telescope. He asked Kepler to continue to show good will toward him. Throughout 1610 and 1611, Kepler did most of the letter writing, and again the correspondence came to a halt. Galileo does not seem to have supplied Kepler with the requested observations and does not mention the telescope nor telescope making. Perhaps Galileo wanted to sell the telescope rather than use it strictly for scientific purposes. He did approach authorities with the suggestion that the telescope could have military uses. Galileo does not credit Kepler and vice versa in any of their publications; neither gave the other a citation.

Galileo did know about the meaning of priority. In 1606, Galileo had published a small book called *The Operations of the Geometrical and Military Compass*. A few months later, Baldessar Capra published a book about the same instrument and claimed the invention for himself and his teacher. On April 9, 1607, Galileo filed legal action against Capra accusing him of plagiarism; he also published a polemic against Capra.

By Newton's time, establishing credit became important for commercial reasons, if not for scientific reputations; patents were issued to discoverers with priority. Newton was involved in several priority quarrels, but he was fortunate in that others did the arguing for him.

In August 1683, Edmund Halley, of comet fame, visited Newton with respect to a discussion he had had with architect

Christopher Wren and Robert Hooke. Halley believed that Kepler's harmonic law relating time of revolution around the sun and distance of planets from the sun could be accounted for by a force between the sun and the planets which diminished as the square of the distance from the sun. Halley could not show this mathematically. Hooke said he had already solved the problem and that all celestial motions could be explained by the law. Christopher Wren offered a "book of forty shillings" to whichever of them could bring proof within two months. Hooke said he would withhold publication so that others might struggle to solve the problem, and when they failed, they would respect him more. Halley asked Newton what kind of orbit planets would have if gravity diminished inversely as the square of the distance. Without hesitation Newton replied, an ellipse. Impressed, Halley urged Newton to collect his lecture notes on gravity, expand them, and publish them. This was the origin of the *Principia,* published in 1687 with funds supplied by Halley. In 1674, Hooke had already published the principle that all celestial bodies had a gravitational attraction toward their centers, and that gravitational attraction fell off with distance according to some law he did not then know. Hooke's work was ignored by Newton.

The two scientists had been adversaries from the start. Seven years younger than Hooke, Newton was a comparative unknown when Hooke had published his *Micrographia* in 1665. The book was a collection of many first achievements in such sciences as biology, meteorology, and optics. It described the first practical compound microscope; it pointed out that a rapid fall in a barometer presaged storms; it had a modern view of fossils. Hooke had become the curator of the Royal Society a few years earlier, and the *Micrographia* also contained a record of some of his experiments. Newton was elected a fellow of the Royal Society in 1672, when he contributed his first paper on light. This was criticized by Hooke, and Newton, hypersensitive to criticism, resigned his fellowship in 1673. In 1679, Hooke asked Newton if he had anything to communicate to the Royal Society. Newton replied that he had just returned from visiting his home village and had been busy with family affairs; nonetheless, he sent Hooke a suggestion for an experiment to show that the earth was rotating. This contained an error and Hooke pointed it out publicly. Later, when Newton was much more famous than Hooke, Newton refused to be president of the Royal Society as long as Hooke was its secretary.

Hooke was also involved in a priority dispute with Henry Oldenburg. In March 1674, Oldenburg had published in the *Philosophical Transactions* a description of Christian Huyghens' recent invention of a watch with a spiral spring applied to the balance to regulate the movement. Oldenburg had a monetary interest in the invention; Huyghens had offered him English patent rights. In anger, Hooke claimed he had made such a watch first, seventeen years earlier. Hooke charged Oldenburg with knowing this fact, and the latter replied that none of Hooke's watches had been successful. Many in the Royal Society supported Hooke, but a majority of the council backed Oldenburg. In November 1676, the council approved a statement published in the *Philosophical Transactions* that disavowed Hooke's contentions. Hooke resolved to leave the Royal Society, but Oldenburg died in September 1677 and Hooke was immediately appointed one of the new secretaries.

At about the same time, another quest for credit caused alienation and broken friendship between thinkers in France. Pierre-Louis Moreau de Maupertuis, who first discerned the true shape of the earth when he led an expedition to Lapland, was one of the first Frenchmen to accept the work of Newton; as a result, he had disputes with the followers of Descartes. The philosopher and man of affairs Voltaire was also an early advocate of Newtonian principles. Voltaire and de Maupertuis were friends. For reasons that are not yet clear, there was an estrangement between the two. Voltaire ridiculed de Maupertuis in writing; *Micromégas* contains one of these adverse criticisms. Voltaire sent a former student of de Maupertuis, Samuel Koenig, to the Berlin Academy of Sciences with a letter to show that de Maupertuis had stolen the principle of least action from Leibniz; the Academy, not being influenced, expelled Koenig. De Maupertuis, on the other hand, claimed that Fermat's principle of least time was actually a corollary of his principle of least action, the idea that natural processes take the path of minimum energy. It may be that de Maupertuis is deserving of more credit in the realm of biology. He anticipated almost every idea of genetics as postulated by Mendel as well as the mechanism for organic evolution adopted in modern times.

As science developed in the nineteenth century, the undue drive for credit, prestige, and honor continued unabated. Even those who were already held in great esteem by their peers and by the public engaged in the activity. By 1823, Humphry Davy was already at the pinnacle of success; he had arrived. Yet his relationship with his protégé Michael Faraday had its ugly moments. In 1823, Faraday was proposed as a fellow of the Royal Society, and Davy vehemently opposed this. Davy went so far as to advise Faraday to have his sponsor, Wollaston, withdraw the candidacy. Faraday refused on the grounds that he had not proposed the nomination in the first place. Davy said that he, as president of the Royal Society, would take all necessary measures to remove Faraday's name. Nonetheless, Faraday was elected, and only one negative vote was cast. Perhaps Davy's opposition was that of a man who wanted to do all the sponsoring of his protégé. But it also could have been jealousy. When European scientists had been asked what they had thought of young Faraday when he had toured Europe with Davy and Lady Davy, they had replied that they had liked Davy but had loved Faraday.

At about the middle of the century, there occurred an argument about the priority of discovery of the law of conservation of energy. The idea can be traced to many people. Between 1842 and 1847, the concept was announced by four different scientists, and even before this the germ of the hypothesis was in the minds of many French, German, and English scientists. Yet Helmholtz allowed himself to be called the father of the law of conservation of energy, and James Joule wrote scientist-popularizer John Tyndall that, he, Joule, was deserving of the credit.

Even generous Michael Faraday once indulged in the drive for credit. He seemed to have a prejudice against William Sturgeon, who also had risen from the lower classes and is usually assigned credit for the invention of the electromagnet and the electromagnetic commutator. Faraday always refused to acknowledge Sturgeon as the inventor, and claimed credit for himself.

The phenomenon of abnormal selfishness in the matter of credit occurred also in the United States. A case scientists would sooner forget occurred between two nineteenth-century paleontologists, Edwin Drinker Cope of the University of Pennsylvania and Othaniel Charles Marsh of Yale University. Marsh, Cope, and Cope's teacher Joseph Leidy were independent collectors of vertebrate fossils in Wyoming. (At that time, scientific expeditions had a military escort to protect the members against Indians.) All three scientists found specimens of an extinct group of fossil mammals called uintatheres, large vertebrates with small brains and three sets of horns in the skull. Leidy was the first to establish priority with a letter dated August 1, 1872. Later in the month Cope made a record of his discovery, and Marsh did likewise. The bitter battle was touched off by Cope's mistakes and Marsh's tactless correction of them. Cope called the tusks of the animals incisors rather than canines and believed the animals were a kind of elephant. The *American Naturalist* printed one of Marsh's attacks; he wrote: "Cope had endeavored to secure priority by sharp practice, and failed. For this kind of sharp practice in science, Professor Cope is almost as well known as he is for the number and magnitude of his blunders . . . Professor Cope's errors will continue to invite correction, but these, like his blunders, are hydra-headed, and life is really too short to spend valuable time in such an ungracious task, especially in the present case. Professor Cope has not even returned thanks for the correction of nearly half a hundred errors . . ." Both Cope and Marsh gave different scientific names to the fossils and both ignored the work of Leidy.

During the next few years the United States government organized and financed geological surveys of the West. Cope was the vertebrate paleontologist for two of these, while Marsh acted in the same capacity for the other two. But later all four surveys were unified into the United States Geological Survey, and Marsh's friends came to head that group.

On January 12, 1890, Cope's views were given space in the *New York Herald*. Cope claimed that Marsh and John W. Powell, the director of the unified States Geological Survey, were "partners in incompetence, ignorance, and plagiarism."

He charged that some of Marsh's work had been taken from others or done by his assistants and that one of Marsh's scientific papers was "the most remarkable collection of errors and ignorance of anatomy . . . ever displayed."

As for Marsh's genealogy of the horse, Cope declared this to be the work of the Russian paleontologist Kòwalevsky. Cope also gave the newspaper letters written to him from others who similarly condemned Marsh. One scientist had written about Marsh: "During most of my time while in his employ, I never knew him to do two consecutive honest days' work in science. . . . The larger part of the papers published since my connection with him in 1878 have been either the work or the actual language of his assistants. . . . He has never been known to tell the truth when a falsehood would serve the purpose as well."

Cope considered his statements about Marsh to be objective, scientific reports because he said after the newspaper articles were printed, "I refuse utterly to have my criticism of Professor Marsh put on the low ground of a personal quarrel."

Marsh's reply was printed in the January 19, 1890, issue of the *New York Herald*. He of course denied each accusation made by Cope and supported the denials with evidence. He also charged Cope with stealing specimens and with plagiarism. He wrote: "During the recent International Geological Congress in London . . . the cases of Cope and Kòwalevsky were fully discussed. . . . Kòwalevsky was at least stricken with remorse and ended his unfortunate career by blowing out his own brains. Cope still lives, unrepentant."

Cope reiterated his accusations in the January 20, 1890, issue of the *New York Herald* and they were printed too in the pages of the *American Naturalist,* owned by Cope. Marsh's former assistants George Baur and Erwin H. Barbour also attacked him in letters to the magazine. The latter assistant called Marsh an incompetent "scheming demagogue" who used trickery and plagiarism.

At the height of the quarrel, Leidy withdrew from vertebrate paleontology, saying that it was no longer a field for gentlemen. He along with Cope and Marsh are acknowledged founders of American vertebrate paleontology.

The abnormal selfishness in the drive for credit has by no means disappeared in the modern day. C. P. Snow, in his *Science and Government,* has documented the controversy between English scientists Tizard and Lindemann during World War II. In 1958, the American Physical Society launched a bimonthly journal, *Physical Review Letters,* in order to have more rapid communication and to establish priority claims more justly; this journal made it possible for the elapsed time between the submission of a manuscript and its publication to be as little as four weeks. In one of the early issues, the editor commented that more than 40 percent of the manuscripts received were not worthy of publication, and added: "We do not take kindly to attempts to pressure us into accepting letters by misrepresentation, gamesmanship, and jungle tactics, which we have experienced to some (fortunately small) extent." Evidently, scientific honesty does not seem to be present for some scientists seeking recognition.

A sociologist studying the scientific community in the early 1960's reported that about 10 percent of his sample of scientists admitted, without his being interested in the question, to having been involved in questions of priority. In the 1968 OECD report analyzing American science policy, biologist C. H. Waddington reported finding much distrust and suspicion among researchers. He wrote: "It is not uncommon to find a scientist in a rapidly advancing field, unwilling to disclose his results in private conversation or at a seminar until they have been safely rushed into print, and he can no longer be scooped by his rivals."

Perhaps several ways are needed to channel the abnormal drive for credit. Mass-production techniques in scientific research seem to have reduced somewhat the individual controversies, but group controversies have been magnified. It was group controversy that was in evidence in some circles of the American atomic bomb project. Groups at Columbia University would, apparently in jest, claim more priority and importance than groups at Oak Ridge, the University of Chicago, or one of the other installations.

What is desirable in the matter of assigning or claiming credit is the wide application of the rules of English sportsmanship that governed the Darwin-Wallace incident. Both arrived at an accumulation of evidence and a mechanism for the theory of organic evolution at about the same time. Instead of indulging in an ugly controversy, both scientists acted like gentlemen when it came to the assigning of credit. (Yet Darwin never acknowledged the idea of natural selection which he must have read in the work of Edward Blyth.) A similar situation prevailed in the discovery of Neptune. Friends of the scientists involved claimed much for them until Sir John Herschel had the good sense to bring together the principals in the discovery, and amicable results were obtained. The Hertzsprung-Russell diagram for the development of stars also had no ugly priority battle. Another pleasant resolution of priority claim was arrived at for x-rays. It has been established that A. E. Goodspeed of Philadelphia had actually, however accidentally, made an x-ray photograph on February 22, 1890, six years before W. K. Roentgen's discovery. Goodspeed said to a University of Pennsylvania audience after the announcement by Roentgen: "We can claim no merit for the discovery, for no discovery was made. All we ask is that you remember, gentlemen, that six years ago, day for day, the first picture in the world of cathodic rays was taken in the physical laboratory of the University of Pennsylvania."

D. Mendeleef, given credit for the nineteenth-century achievement of the periodic table of chemical elements, said: "Science is everybody's property, and consequently, justice demands to render the utmost scientific glory not the first sayer of a known truth but the man who first persuaded others of it, showed its probity, and made it scientifically admissable. Scientific discoveries are seldom made at once; usually their messengers are short of time to prove the truth of the discovered; time calls into being a real creator possessing all means to render truth everybody's conviction; one must not, however, forget that this creator can appear only due to the work of many others and to the accumulated store of data." Others did have a periodic table at this time, but Mendeleef had the courage to predict the properties of new elements with his table; three were found in his lifetime with almost the same properties as he had predicted.

If scientists would look at the record, then the end should be near for the nonsense of priority struggles. In 1922, sociologists William Fielding Ogburn and William Isaac Thomas published a list of 148 cases wherein scientific discovery could be justifiably credited to more than one source. In 1961, sociologists Robert Merton and Elinor Barber studied 264 such cases, and 179 of them had two simultaneous discoverers.

Scientists have made some suggestions to alleviate the problem of assigning priority. In the January 7, 1966, issue of *Science,* Irvine H. Page of the Cleveland Clinic urged the establishment of such rules as that data must be presented in a paper that is published in a journal which has reasonable editorial supervision and that the date of receipt and the date of acceptance of the manuscript are printed with it. Late in 1966, Dr. Samuel Goudsmit, editor of *Physical Review Letters,* suggested that precise reviews by highly competent authors would solve the problem of information retrieval, and scientists would no longer be burdened with struggling through large numbers of papers.

All of the suggested techniques to diminish the abnormal drive for recognition will end in failure unless science education inculcates in the student a commitment to the subject and to mankind, rather than a commitment to self-enhancement. Perhaps a written accepted dictum comparable to the Hippocratic Oath in medicine would have a sobering effect. There exists in the history of science enough incidents of cooperation to embellish an oath that may be acceptable.

The manner in which science progresses, developed in the remaining chapters, needs also to be a part of the education of the scientist.

SELECTED REFERENCES

Merton, R. K. "Priorities in Scientific Discovery," *American Sociological Review,* 22 (1957), 635–59.
———. "Resistance to the Systematic Study of Multiple Discoveries in Science," *European Journal of Sociology,* 4 (1963), 272–82.

Reif, F. "The Competitive World of the Pure Scientist," *Science,* 134 (December 15, 1961), 1957–62.

Snow, C. P. *Science and Government.* New York: Mentor Books, 1962.

Wheeler, W. H. "The Uintatheres and the Cope-Marsh War," *Science,* 131 (April 22, 1960), 1171–76.

12
THE PROGRESS
OF SCIENCE

One way to avoid the selfish drive for credit among scientists is to have science students study the history of science, with emphasis on its continual change of theory and on the idea that every generation of scientists builds on and adds to the knowledge accumulated by its predecessors. The founder of nuclear science, Ernest Rutherford, said it this way: "It is not in the nature of things for any one man to make a sudden violent discovery; science goes step by step, and every man depends on the work of his predecessors. . . . Scientists are not dependent on the idea of a single man, but on the combined wisdom of thousands of men." The late nineteenth- and early twentieth-century American physicist S. P. Langley described the progress of science metaphorically as the activity of a pack of hunting dogs "which in the long run, perhaps, catches its game, but where nevertheless when at fault, each individual goes his own way, by scent, not by sight, some running back and some forward, where the louder-voiced bring many to follow them, nearly as often in the wrong path as in the right ones; where the entire pack even has been known to move off bodily on a false scent."

The idea can be appreciated too in the development of scientific ideas. Early in the intellectual history of man, earth, fire, water, and air were considered the prime elements of nature. After a few hundred years of use of this classification, a fifth element, quintessence, was added. Mercury, salt, and sulfur were the basic units for the early chemists, and after some generations, the idea of chemical element took hold. Until the nineteenth century, caloric was an imponderable used to interpret heat energy, but at the end of the century, the concept of molecule for heat energy interpretation began to dominate. This was the time too when the ether was dropped, and the fundamental particles in vogue today began to be adopted. In astronomy today, the largest diameter of our galaxy is said to be 100,000 light years, but in 1915, it was a different figure. The distance to the Andromeda galaxy was stated as 800,000 light years in 1940, while today it is considered to be about 2,000,000 light years distant. It may be that the increased tempo of scientific activity that characterizes our times will speed up the revisions which have previously taken generations.

Technology too has a record of revision through the generations. The process for making sodium carbonate, soda ash, devised by Nicholas LeBlanc during Napoleonic times, was to heat salt and sulfuric acid and the sodium sulfate formed was reacted with soda and limestone. During the early twentieth century, the Solvay process was devised; salt, ammonia, and carbon dioxide are heated to form sodium bicarbonate, which is then decomposed to soda ash. Prior to the twentieth century, coal had been burned to form coke, and all other by-products were discarded. By-product coke ovens were introduced at the beginning of the century and have resulted in the production of many valuable chemicals. One of the first so produced, indigo, had a great influence on madder plantations; the madder plant is the natural source of indigo. Prior to the middle of the twentieth century, steelmakers used a blast of air rather than one of oxygen; coal was not heated under pressure to produce such products as phenol, naphthalene, and anthracene. Benzene was not produced commercially from petroleum until 1950; in 1960, two thirds of the total United States production of benzene came from petroleum.

Many current ideas in science and technology started out with very few adherents. Their wide acceptance today tempts agreement with philosopher Schopenhauer that each truth enjoys but a short life-span of interested acknowledgment that lies between being ridiculed as absurd and discarded as trivial.

A case history that could be used for the instruction of all scientists and engineers is the development of our solar-system concept. It illustrates very well the theme that the progress of science depends upon several generations of workers.

At the beginning of the sixteenth century, the intelligent man was a firm believer in the geocentric conception of the solar system. He knew that the earth was at the center of a system of objects clearly seen moving about him. The moon made its monthly cycle in twenty-seven and one-third days; in that time the moon was in the same place at the same time against the background of stars. The sun could be seen in the same place among the stars at the same time only once a year; the sun took one year to go about the earth. The so-called fixed stars turned once every twenty-four hours.

At the beginning of the sixteenth century, the intelligent man knew the regular motions in the sky. Diurnal motion, the turn of the sky once every twenty-four hours, was easily discernible. The sun made this turn during the day, and the imaginary concentric circles made by the stars about Polaris, the North Pole star, were its manifestation at night. Almost from childhood, everyone knew about the eastward drift of the sun, moon, and planets. In Europe, these objects are seen in the southern skies. Each twenty-four-hour period, the moon moves farther east of its previous position. The sun and the planets do the same, but at a much slower rate. Intelligent observers also saw the occasional but regular wandering of the planets to the west. This retrograde motion prompted the earliest Greeks to call the objects wanderers, or planets.

Diurnal motion, eastward drift, and retrograde motion were explained by the geocentric conception. The stars as well as everything in the sky turned once every twenty-four hours; this was diurnal motion. The sun, moon, and the planets drifted farther east because they traveled so much slower about the earth than did the stars. The retrograde motion of the planets, claimed the theory, resulted from their manner of encircling the earth. The planets did not have simple circular paths about the earth; they traveled in smaller circles, epicycles, the centers of which could be connected to form a large circle about the earth called the deferent.

The geocentric conception was systematized by the Alexandrian scientist Claudius Ptolemy. He developed it so well that for many it was the Ptolemaic idea—evident, accepted, and established. It found many adherents and lingered as the truth for many centuries.

From the time of its inception until about the eighteenth century, the Ptolemaic system can be considered a first approximation of the truth. As an approximation, it was corrected by a series of advances. This is characteristic of the method of advance in every science. Each successive generation of scientists modifies a conception until finally a new model replaces the old. In some cases, the process has been long and drawn-out, replete with bitterness and personal attacks. The twentieth-century German physicist Max Planck said that ideas change in science not because logic and evidence are against them but rather because an old generation of scientists dies and a new generation with unprejudiced minds appears.

The geocentric view had its competitors in early Greece. In both the third and fourth centuries B.C., the opposing idea of a sun-centered universe was stated. Aristarchus of Samos may have echoed Heraclides of Pontos, but neither time did the idea take hold because simple observation could not support the theme.

At the beginning of the sixteenth century, the Ptolemaic system with man and his planet at the center of the universe had vexing problems to explain. Closer observations of the planets and of star positions demanded adjustments; small changes in revolution times were made and more epicycles were added. For some imbued with the ancient, medieval, and modern value of simplicity, the epicycles became a cumbersome necessity. Thinkers were provoked.

Nicholas Copernicus, a German-Polish cleric, was one so disturbed. He may have been stimulated by the Church's need for a better calendar. His teachers could have been the influence that set him to the problem. He had an extensive university education at some of the best schools in Europe; one of his professors did have the reputation of being immensely dissatisfied with the Ptolemaic view. Young Copernicus studied at Krakow University in Poland, in Padua and Bologna in Italy, and at the University of Ferrara in Italy, where he obtained a doctorate of canon law degree. Old astronomical instruments used by Copernicus as a student at Krakow are preserved today in a museum there.

Copernicus said he read about a competing idea in the works of Aristarchus of Samos. Having an excellent position in the Church hierarchy, Copernicus had the leisure to develop the idea. At the same time, he had the leisure to be a poet, economist, physician, and philosopher. In economics, he worked out a system of monetary reform and stated the principle that debased money tends to drive out good money, an idea also attributed to Sir Thomas Gresham and known as Gresham's law. He left a number of interesting medical notes. He painted his own portrait and published a translation into Latin of eighty-five Greek verses by a Byzantine poet named Theophylactus Simocatta, who lived in the seventh century.

The experimental or observational point of view was not then in vogue, and Copernicus did not feel compelled to substantiate his thought with observation; he made only a few observations. He developed the Copernican system over a period of years. He was not beleaguered by reports to supervisors, appearances before committees, or the urge to publish for whatever emoluments accrue. He first circulated his work in a less-complete form before he was ready to present it fully. The concept became clearly defined in his mind in about 1510, as indicated in copies of a treatise unpublished in his lifetime, *Commentariolus* (Little Commentary). A reference regarding the development of his theory is in the letter written to his friend the Krakow geographer Bernard Wapowski in 1524.

By the time his great work was ready, Copernicus was on his death bed. The result is that the introduction to his book *De Revolutionibus Orbium Coelestium Libri VI* (Six Books Concerning the Revolution of the Heavenly Spheres), was written by another. George Joachim Rheticus, professor of mathematics at Wittenburg University, made the first public announcement of the Copernican theory in the Baltic city of Gdansk in 1540. First praise for the theory is recorded at Krakow University in 1542. However, the Ptolemaic theorists were strongly entrenched, and a number of universities officially denounced the Copernican theory—Wittenberg in 1542, Zurich in 1559, Rostock in 1573, Heidelberg in 1582, Tübingen in 1586, and Jena in 1593. However, the Copernican concepts were taught at Krakow University in 1549, the University of Salamanca in 1561, and Oxford University in 1576.

Some modern scholars view the achievements of Copernicus with tempered enthusiasm. They claim it is the work of an ancient, that it is encumbered by Ptolemaic epicycles as well as the mathematical tools employed by the ancients. Copernicus only reduced the number of epicycles; he did not do away with them. The more popular interpretation of the worth of Copernicus is that he was a great scientist who had the courage to place the sun at the center of the solar system.

Among the first to accumulate the necessary facts to support Copernicus was the Danish nobleman Tycho Brahe, sometimes called the astronomer with a brass nose because part of his nose had been sliced off in a youthful duel and the part missing had been replaced with metal. Brahe had the unique position of being an astronomer on an island especially designated as his; it was complete with lavish buildings and equipment. No astronomer before or since ever had such total support; no astronomer before ever had the opportunity to devote all his time to the subject. With the help of many assistants and with an entourage befitting a king, including a court jester, Tycho Brahe collected a great amount of information. His good deal came to an end soon after the succession of a new king of Denmark. Brahe then moved to Bohemia, where he secured the support of Rudolf II.

Brahe was not too interested in theory. Nonetheless, he developed a compromise between the Copernican and Ptolemaic views that came to be known as the Tychonic system. Its worth is indicated by its longevity; it did not last longer than the lifetime of Tycho Brahe.

While in Bohemia, he had an outstanding German genius as his assistant. Johannes Kepler, a student of theology until the age of twenty-four, had impressed many scholars and had shown a special talent for mathematics. Brahe was delighted to have him, although their preliminary negotiations as well as later disputes about terms of employment would seem to indicate a contrary attitude.

Johannes Kepler is similar to practically all the pioneers of modern science in that he had religious training and background, but he is one of the few with peasant origins. He was born seven and one-half months after Heinrich Kepler had married Katherine Guldermann. Kepler later described his father, a mercenary soldier when employed, as "vicious, inflexible, and quarrelsome" and his mother as having "a bad disposition." Johannes was a sickly child with poor vision, skin ailments, and digestive troubles. After a preliminary education, he attended a theological seminary. He graduated from the University of Tübingen, and continued to study four more years in the theology division. At the age of twenty-four, he was unexpectedly offered a position teaching astronomy and mathematics in Gratz, the capital of the Austrian province of Styria. He had few students in his class, and the school authorities excused this "because the study of mathematics is not every man's affair." He cast horoscopes as part of his duties, and although he did not care for this work, he was a defender of astrology. At one time he wrote that "while justly rejecting the stargazer's superstitions, they should not throw out the child with the bath water." He believed that the moon had inhabitants and that comets had will and purpose, like fishes in a sea.

Kepler wrote his first book when he was twenty-five years old. In his *Mysterium Cosmographicum,* he presented the idea that five perfect solids could be constructed in the spaces between the planets. The solids are the cube, tetrahedron, octahedron, dodecahedron, and icosohedron; they are perfect solids in having faces of four squares, four equilateral triangles, eight equilateral triangles, twelve pentagons, and twenty equilateral triangles, respectively. He said that the orbit of Saturn could be inscribed in a cube while that of Jupiter circumscribed the cube. The orbit of Jupiter could be inscribed in a tetrahedron while that of Mars circumscribed this figure. This process went on to Mercury. The idea gave Kepler justification for his belief that there were only six planets and that God was a geometer. (During the middle of the twentieth century, the structure of viruses was interpreted with the same five regular solids.) The idea was not quite correct and Kepler rationalized away some of the discrepancies. For Jupiter, he said, "nobody will wonder at it, considering this great distance." For Mercury, he inscribed the sphere containing its orbit not on the faces of the solid but on its edges.

Kepler's major publication was prepared after he succeeded Tycho Brahe as imperial mathematician at the court of Rudlof II. For little more than ten years, Kepler was supported by the monarch and free enough from economic difficulties to concentrate on his work. His *New Astronomy,* published in 1609, contains two of his three laws of planetary motion. The third, the harmonic law, came later. The three laws of planetary motion are affirmed today, and they apply as well to a system of one star going about another and to an artificial satellite going about the earth.

Kepler's first law corrects Copernicus, who said that the planets go about the sun in circular orbits. The first law of Kepler describes the orbits as elliptical, flattened circles. A circle has one point, the center, and one line, the radius, defining the size. An ellipse has two points, the foci, and the distance between them defines the degree of flatness of the ellipse; the farther apart the two points, the flatter is the ellipse. In planetary orbits the two foci are quite close and the sun is at one of the two points.

Kepler's second law describes the amount of space between any two lines drawn from one focus to the ellipse perimeter. He said that the lines, the radius vectors, sweep out equal areas in equal times. Thus, the space bounded by the three points—the earth in its orbit on January 1, the earth in its orbit on January 30, and a focus—is equal in area to any other thirty-day interval. The second law, as well as the third, indicates that a planet travels fastest when closest to the sun.

Kepler had to work another ten years before he arrived at his third law, the harmonic law. It relates the time it takes a planet to go around the sun and its average distance to the sun. The square of the time of revolution divided by the cube of the mean distance is a constant.

Kepler's contemporary Galileo helped the Copernican system along with his telescopic observations. He saw features of the moon clearly enough to decipher the craters. The dark spots on the sun that grow in size and number periodically became visible to him. He clearly saw four moons of Jupiter. He saw Venus in the same kinds of phases in which we see our moon. Galileo made the mistake of thinking that Saturn had four moons because he saw the rings of Saturn at such a tilt, observable about every fifteen years, that only the extreme points of the rings were seen, and these looked like moons. Galileo's pen was even more potent than his observations; he eloquently presented the Copernican view.

Isaac Newton was born in the year that Galileo died. Newton's laws of motion and universal law of gravitation gave the heliocentric view the pillar of support that was needed. Rational explanations were available as a result of Newton's principles. For example, the planets and the sun attracted each other in accordance with the natural force that existed between any two bodies; this force of gravitation is governed by the mass of the bodies and the inverse square of the distance between them. This force would have planets and sun crash into each other, but planets and all other bodies have a natural tendency to move uniformly in a straight line. Were the planets and the sun perfectly homogeneous, spherical bodies, the planets would, as a kind of compromise between the tendency to go into the sun and the tendency to move uniformly in a straight line, go in a circular orbit about the sun. But the planets and sun are not homogeneous spheres; the planets also attract each other, and the moons, meteors, and comets interfere to make elliptical orbits. The principles of Newton enabled the average intelligent person to have an explanation of questions plaguing him: Things did not fall off a rotating earth because they were strongly attracted to the earth.

Newton and the scientists of his time had no means to answer the complaint that great winds would be directed and generated by a rotating earth. As the earth was explored, the regional winds were more completely charted and their responsiveness to earth rotation was ascertained. Newton correctly answered the charge that if the earth turned about the sun, the near stars would be seen in different perspectives, in time, against the background of the more distant stars. He said that the stars were too far away to discern such stellar parallax. He was right, but an entire century elapsed after Newton before the first stellar parallax was measured.

Newton's synthesis bolstered the heliocentric view. It showed too that astronomical phenomena can be correctly interpreted with terrestrially found laws; that the quantitative science introduced in the medieval Parisian universities and accented by Kepler and Galileo, among others, helped find the correct answers.

SELECTED REFERENCES

Goran, M. "The Quiet Revolutionists," *Science Education,* 50 (October, 1966), 335–36.

Hall, A. R. *From Galileo to Newton, 1630–1720.* New York: Harper & Row, 1963.

Hall, M. B. *The Scientific Renaissance, 1450–1630.* New York: Harper & Row, 1962.

Koestler, A. *The Sleepwalkers.* New York: Macmillan, 1959.

Kuhn, T. *The Structure of Scientific Revolutions.* Chicago: University of Chicago Press, 1962.

13
DIALOGUE

In the early eighteenth century, Dr. Johann Bartholomew Adam Beringer, a doctor of philosophy and of medicine, was a senior professor, and dean of the medical faculty at the University of Würzburg, and the adviser to the chief physician of the Prince and Bishop of Würzburg. As a pastime, Beringer busied himself in the study of oryctics, or things dug from the earth.

Beringer approached his avocation of studying stones rationally, avoiding the interpretations of the ancients who did not experiment and who assigned many of the forces of nature to occultism; he similarly scored the astrologers. In true scientific manner, pioneering for a man who lived in the seventeenth and eighteenth centuries, he critically surveyed the immediate past as well as the then held opinions about the nature of the stones he found. Some of the ancient thoughts discussed by Beringer in his treatise seem strange to a modern man. Today, scientists do not assign the origin of fossils to the action of the stars, to a process allied to fermentation while the rocks were still soft and plastic, to a practical joke by nature, to practice by the great architect of the universe for producing the real thing, to the evil powers wishing to deceive mankind, to the seeds of the mineral world corresponding to the seeds of the animal and vegetable world.

Beringer employed three boys, aged eighteen, seventeen, and fourteen, to help him gather "figured stones." After examining the stones, he claimed that they were unique because they had letters on them that were raised and prominent. Unfortunately, his "Würzburg stones" were not scientifically unique at all; they were part of a cruel hoax being played on him.

Two of his colleagues, J. Ignatz Roderick, professor of geography, algebra, and analysis at the University of Würzburg, and Georg von Eckhart, Privy Councilor and Librarian to the Court and the University, had hit on a plan to dupe Beringer. It is not known whether they did it out of revenge or envy (according to them, Beringer was arrogant and contemptuous of them) or as a practical joke (which grew to proportions beyond those they had planned).

Whatever the cause of it, the hoax was carried out according to plan. Roderick and von Eckhart hired one of the boys who worked for Beringer, Christian Zänger, to take part in the trickery. Roderick carved figures into fragments of shell limestone, and Zänger polished the "stones." The boy then placed these stones where they would be "found" by the other boys, the Hahn brothers. Zänger also brought some of the manufactured stones to Beringer.

Beringer published his findings concerning the stones in his *Lithographiae Wirceburgensis*. Soon after, the professors involved in the hoax decided to bring it to an end. But instead of informing Beringer directly that he had been duped, they started a rumor that the stones were fraudulent. (The popular version of the climax is that Beringer found his name on one of the stones and awoke to the realization that he had been fooled, but there is not substantiation for this version.)

Beringer immediately sought to clear his name, but the damage was done. He lived for fourteen more years and wrote two more books. But his *Lithographiae Wirceburgensis,* containing accounts and diagrams of the false stones, became the basis of his reputation. He did attempt to purchase all the copies of the book, but it was too late. Many were circulated beyond his reach and are still available as a collector's item.

The Beringer incident is not an isolated case. There are several comparable cases in the annals of biology, geology, anthropology, and archeology. On May 14, 1864, a meteorite fell near the village of Orgueil, near Toulouse, in southwest France. Twenty fragments were collected and stored in museums; all were of a variety called carbonaceous chondrites—soft, dark, readily soluble in water, and suggestive of extraterrestrial life. One hundred years later, during a scientific dispute about these meteorites, some were opened and analyzed. Scientists were astounded to find fragments of a plant later identified as a European weed, together with gravel, fragments of coal, and glue within the meteorite. Evidently, someone in 1864 had deliberately placed organic material in the specimen. Perhaps it was a sordid joke. Or it may have been done in reaction to Louis Pasteur's then passionate defense of divine creation as the only origin of life.

At Cardiff, New York, in 1889, a planted figure twelve feet long was called undoubtedly ancient by responsible scientists until others more analytical than they scoffed at the evaluations. The Cardiff giant did earn money for its perpetrators, who charged admission to see it, and for the American showman P. T. Barnum, who made a replica of it.

Early in the twentieth century, part of a skull was found at Piltdown, England. At the time some scientists thought it was the fossil of an early man—*Eoanthropus*. But subsequent scientific investigation showed that it had been constructed in an elegant manner from two separate pieces, a definite attempt to deceive. In 1928, artifacts found near Vichy, France, caused many arguments among amateur and professional archeologists. Some contested the authenticity because of the condition of the materials and were vindicated by chemical and microscopic examinations. Even in 1969 a Smithsonian Institution expert on primates wanted the Federal Bureau of Investigation to x-ray an attraction at midwestern carnivals known as "The Minnesota Iceman." The phenomenon had been noted in a scientific paper by a Brussels University scientist. Eventually the operator of a wax museum in California said the "previously unknown life form" resembling a human male had been made from rubber and hair by his technicians.

Hoaxes in science occur because scientists, like other people, may engage in practical jokes. Such behavior can be considered an extreme example of interaction between scientists. Unfortunately, hoaxes are not the only kind of unwelcome and unnecessary interactions.

In 1877 one example of professional exchange in the science

of chemistry concerned organic chemist Hermann Kolbe who, in writing about van't Hoff's idea of three-dimensional arrangements of atoms in molecules, said, "Will anyone to whom my worries may seem exaggerated please read, if he can, a recent memoir by a Herr van't Hoff on 'The Arrangements of Atoms in Space,' a document crammed to the hilt with the outpourings of a childish fantasy. This Dr. J. H. van't Hoff, employed by the Veterinary College at Utrecht, has, so it seems, no taste for accurate chemical research. He finds it more convenient to mount his Pegasus (evidently taken from the stables of the Veterinary College) and to announce how, on his daring flight to Mount Parnassus, he saw the atoms arranged in space." That Kolbe was himself in error seems to be substantiated by the fact that van't Hoff's idea had also been independently stated by French chemist Julius Achille Le Bel and was soon being adopted by other chemists, and by the fact that van't Hoff was the first recipient of the Nobel prize in chemistry in 1900.

Dialogue at the lower levels of scientific debate continues today. In 1959, two experts on Galileo reviewing Arthur Koestler's *The Sleepwalkers* for the scholarly journal *Isis,* the official quarterly journal of the History of Science Society, used such language as "insolent misrepresentation," "deliberate distortions," and "dishonest." In the November 1964 issue of the *Scientific American,* a psychologist-letter writer called a sociologist-author ignorant of one field and lacking in understanding of another.

The extremes of hoax and calumny in dialogue are inevitable. But they need not be as frequent if undergraduate students, graduate students, and scientists are made aware that science lives on friendly controversy. Philosopher Hegel said that synthesis results from thesis and antithesis; philosopher Dewey claimed that growth comes with such interactions as dialogue.

In the twentieth century, scientists and science students have been largely deprived of the benefits of dialogue. Earlier, debates between opposing schools of thought, controversy in the journals, and criticism at meetings were the rule; now they are the exception. This is how physicist Arnold Sommerfeld viewed the educational experience of scientific debate: "The champion

for energetics was Helm; behind him stood Ostwald, and behind both of them the philosophy of Ernst Mach (who was not
present in person). The opponent was Boltzmann, seconded by
Felix Klein. The battle between Boltzmann and Ostwald was
much like the duel of a bull and a supple bullfighter. However,
this time the bull defeated the toreador in spite of all his agil-
ity. The arguments of Boltzmann struck through. We young
mathematicians were all on Boltzmann's side."

Today's journals of science hardly ever feature friendly controversy. Hot words may be exchanged in their letters columns
about the social, political, and economic implications of science.
The small amount of space given to correspondence may at
times contain short expressions of dissent about scientific material. Albert Szent-Györgi, a Nobel-prize winner, mildly rep-
rimanded a scientist who had claimed Szent-Györgi had been
"unfortunately erroneous" when he wrote on page 400 of the
January 28, 1966, issue of *Science*: "What Ti Li Loo probably
meant is that his findings differed from those of Friedlander and
French."

Today's discussions at seminars feature no more than polite
questions, hardly ever impinging upon basic issues. Because the
speakers are generally experts in some very narrow, specialized
field and because their delivery is frequently uninspiring to
boot, the meetings often do not involve the scientists in an educational manner. The student of science, even if he is fortunate
enough to have research experience, may retain the impression
that science is a matter of authority. Diverse interpretation of
theory is not within his ken.

Today's student may even be conditioned to the unusual
case of a scientific controversy turning to the courts for judgment. In 1970, a British dental surgeon alleged that the *British
Medical Journal* had libelled him and he sued. In 1962 his paper
in the journal had described a new technique of anaesthesia
and he had acquired an international reputation. In 1969 the
journal published a critical report by four scientists. Lord Den-
ning, Master of the Rolls, commented: "It would be a sorry day
if scientists were to be deterred from publishing their findings
for fear of libel actions. So long as they refrained from personal

attacks, they should be free to criticise the systems and techniques of others. Were it otherwise, no scientific journal would be safe."

Big science, a bureaucracy, is subject to the same cold, impersonal forces that pervade society; the dialogue between scientists has been conditioned by these forces. The research worker has available, in general, only the squeezed-out result, minus the developments and side paths that could be instructional to him.

Truly, the final results of scientific research are often enormous and impressive. But a great deal of the research that led up to these results could be useful as case studies for students—the arguments pro and con, the unfruitful excursions, the mistakes, and the misjudgments—all of these could be just as instructive as the final results. For instance, soon after the discovery of x-rays, a French scientist found "N-rays." Before the concept was exposed as a mistake, about one hundred papers were published on the subject in the French journal *Comptes Rendues* in 1904 and the French Academy gave twenty thousand francs and a gold medal to the "discoverer." Eventually experts in other countries showed that "N-rays" do not exist. In 1894, Lord Rayleigh and William Ramsay of England announced they had found a new element, argon, in air. Such an element had not been predicted by the periodic table of chemical elements developed by Mendeleef; there was no place for it on the table. Mendeleef said the new gas was evidently a molecular form of nitrogen. This wrong opinion was forgotten as the properties of argon were found, and the periodic table was modified to make room for it and the other noble gases. Early in the twentieth century, Lord Kelvin questioned the phenomenon of radioactivity; he thought radium was not a chemical element. In rebuttal, Rutherford showed that if radium were a chemical compound containing helium, it was an entirely different kind of compound. Radium released a million times more energy per unit mass than any known chemical change, and its rate of decay was independent of temperature. Niels Bohr, the Danish scientist who applied quantum theory to the structure of the atom, assumed that the electron orbits were circular. About a decade later, Sommerfeld in Germany showed that elliptical orbits as well as circular ones were present. The examples capable of being expanded and used seem to be legion.

Dialogue is weak in applied science and engineering for the same reasons that it is in pure science; an additional factor is that trade secrets and patents are not open for discussion. Here too, the final results may become widely known while the paths leading to them are forsaken and even forgotten by those few who took part. There have been some in which more dialogue could have prevented disastrous accidents. In the January 27, 1967, spacecraft fire that took the lives of three American astronauts, a subcontractor was accused of numerous deficiencies in design, manufacture, installation, and quality control. The only astronaut on the investigating board, the command pilot of Gemini 7, Frank Borman, said that no one "gave serious concern to a fire in the spacecraft. We tried to identify every hazard we could find, but this was one we missed." Extensive dialogue about the possible hazards may have helped. In other situations too, every conceivable difficulty must be checked. In 1967, the National Communicable Disease Center reported that one fourth of the 500,000,000 medical laboratory tests run each year produce incorrect or even fatal results. In 1955, at least four million children were inoculated with a polio vaccine that was contaminated with a virus, SV-40; fortunately, this mistake has had no serious consequences.

Better dialogue in science, although time-consuming, can also be beneficial in communicating with and educating nonscientists. If the latter were made aware of the controversy among scientists, they would no longer view science as a repository of final and absolute truth. They would see that science is everything from an educated guess to the most certain fact. With such an informed public, science would no longer be used in sales and advertising. The white-smocked laboratory worker would be pictured less often supporting the claims of a mouthwash, toothpaste, or deodorant. Intelligent men and women would also have a better appreciation of the part financing plays in the advancement of science; they might even give valuable help in allocating funds to science.

SELECTED REFERENCES

Adams, F. D. *The Birth and Development of the Geological Sciences.* New York: Dover, 1954.

Jahn, M. E., and Woolf, D. J. *The Lying Stones of Dr. Johann Bartholomew Adam Beringer being his Lithographiae Wirceburgensis.* Berkeley: University of California Press, 1963.

MacDougall, C. *Hoaxes.* New York: Dover, 1958.

McKusick, M. *The Davenport Conspiracy.* Iowa City: University of Iowa Press, 1970.

14
SELF-IMPROVEMENT

The so-called knowledge explosion has prompted many commentators to suggest periodic retooling for many professionals. But in the case of research scientists, re-education is already a part of their work. In the task of finding out something new, the scientist first immerses himself in what is already known and then continually criticizes his own efforts. The problems of science education are enormously eased by this advantage.

Every first-rate scientist admits that many of his ideas of how to solve problems never reach the testing stage because of his own adverse criticism. Louis Pasteur advised: "When you believe you have found an important scientific fact and are feverishly curious to publish it, constrain yourself for days, weeks, years sometimes; fight yourself, try to ruin your own experiments, and only proclaim your discovery after having exhausted all contrary hypotheses." Michael Faraday said: "The world little knows how many of the thoughts and theories which passed through the mind of the scientific investigator have been crushed in silence and secrecy by his own severe criticism and adverse examination; that in the most successful instances, not a tenth of the suggestions, the hopes, the wishes, or the preliminary conclusions have been realized." He also said: "I always tried to be very critical of myself before I gave anybody else the opportunity, and even now I think I could say much stronger things against my notions than anybody else has." Humphry Davy confessed, "the most important of my discoveries have been suggested to me by my failures." Johannes Kepler said, "how many detours I had to make, along how many walls

I had to grope in the darkness of my ignorance until I found the door which lets in the light of truth."

The documentation of this kind of self-corrected mistake depends upon the cooperation and the degree of articulation of the scientist. Before he was famous, Albert Einstein applied the quantum theory to the phenomenon of specific heat; he thought only the vibrations of positive ions were involved; recognizing his error several months later, he published a correction to his paper, allowing for vibrations of neutral atoms. When he was a celebrated scientist, Einstein said that the biggest blunder of his life had been the introduction of a cosmic constant into his relativity equations. The eighteenth-century chemist Joseph Priestley documented his false starts and mistakes in his *History of Electricity*.

Probably the best self-education technique in science is the method of multiple hypotheses. T. C. Chamberlin, a geologist at the University of Chicago, described it at the beginning of the century: "The moment one has offered an original investigation for a phenomena which seems satisfactory, that moment affection for his intellectual child springs into existence, and as the explanation grows into a definite theory, his parental affections cluster about his offspring and it grows more and more clear to him . . . There springs up also unwittingly a pressing of the theory to make it fit the facts and a pressing of the facts to make them fit the theory. . . . To avoid this grave danger, the method of multiple working hypotheses is urged. It differs from the single working hypotheses in that it distributes the effort and divides the affections. . . . Each hypothesis suggests its own criteria, its own means of proof, its own method of developing the truth, and if a group of hypotheses encompass the subject on all sides, the total outcome of means and of methods is full and rich."

Perhaps it takes a great scientist to correct himself before or after publication. French scientist Henri Le Chatelier wrote in his *Science et Industrie*: "Men who are capable of modifying their first beliefs are very rare. This ability was one of the reasons for the success of Claude Bernard and Pasteur. Out of a very vivid imagination they forged new hypotheses all the time but abandoned them with equal ease as soon as experience contradicted them." Perhaps more scientists should have the opinion of Roger L. Stevens, chairman of the U.S. National Endowment for the Arts, who said late in 1965: "I always keep raising a rather obvious but often overlooked fact about science. Science consists of a series of mistakes looking for an answer. . . . It is failure after failure after failure."

Since persistence in belief is another important attribute of greatness, the question becomes: when does a scientist discard an idea and when does he maintain his belief? A case study may be revealing.

Humphry Davy was born in the Cornwall section of England on December 17, 1778. His birthplace, Penzance, was the site of the Wherry mine, where one of James Watt's steam-pumping engines, located on land, serviced the shaft of the mine a mile at sea. Cornwall was where the first high-pressure steam engine was built and coal-gas lighting was invented.

Davy's paternal grandfather was a successful builder, while his father, besides operating a small farm, was one of the last commercial wood carvers in the area. Upon the death of her parents, Davy's mother and her two sisters, all very young, had been adopted by a local physician and surgeon. Much of Humphry Davy's childhood was spent in this home.

Davy's early formal education left no lasting mark. He later wrote: ". . . the way in which we are taught Latin and Greek does not much influence the important structure of our minds. I consider it fortunate that I was left much to myself as a child, and put upon no particular plan of study, and that I enjoyed much idleness   I perhaps owe to these circumstances the little talents I have, and their peculiar application . . ."

His informal education may have been more instructive. He was a youthful raconteur and poet, an entertainer with homemade fireworks, a fisherman, a hunter, and a participant in dramatics improvised by him and other youths. The French revolution too may have influenced him. He learned French from refugees from France; one of his boyhood loves was a French girl to whom he wrote many sonnets.

Humphry Davy's father died at the age of forty-eight, in 1794. Shortly thereafter, his mother started a millinery shop in partnership with a French woman who had fled the revolution.

Humphry Davy was apprenticed to a surgeon and apothecary.

Davy had other forces directing him to science. He became acquainted with Davies Giddy, who later changed his name to Gilbert and became a president of the Royal Society. Gilbert took youthful Davy to visit a chemical laboratory and Davy was delighted. James Watt's son, Gregory, came to Penzance to rest for his health and found a room in Mrs. Davy's house. University-educated Gregory Watt and Davy became friends and had spirited discussions about science.

The industrial revolution was one of the greatest spurs in Davy's career. He was aware of this revolution and was eager to participate in the application of science to it. In one of his many letters to Mr. Gilbert, he wrote of his desire to see the steam railroad serving on the roads of England. One of the first series of lectures he later gave in London was on the art of tanning. In another he said, "the man of science and the manufacturer are daily becoming more assimilated to each other." He delivered a series of lectures about the chemistry of agriculture. He wrote, "nations shall know that it is their interest to cultivate science, and that the benevolent philosophy is never separated from the happiness of mankind."

Within two years after being apprenticed, at the age of twenty, Davy was brash enough to publish a theory on the nature of light. His work provoked some scorn. Almost immediately, he referred to this twenty-thousand-word report, "An Essay on Heat, Light, and the Combinations of Light," as his "infant speculations" and did not care to be reminded of it. Two years after this publication, he began his next with a statement explaining that "early experience has taught me the folly of hasty generalization." (During the sixteenth century, Tycho Brahe had divided his work into "childish and doubtful" during his student days, "juvenile and habitually mediocre" upon beginning his career, and finally "virile, precise, and absolutely certain.")

Davy's first report was given to Dr. Thomas Beddoes by Gregory Watt. Beddoes, a physician, had been a professor of chemistry at Oxford University and had translated the *Chemical Essays* of late seventeenth-century Swedish chemist Scheele, the discoverer of chlorine. Beddoes, in Bristol, was impressed with

Davy's essay which contended that light "enters into the composition of a number of substances" and that oxygen gas contains light and should be called phosoxygen. Beddoes had established a Pneumatic Institution to explore the medical possibilities of gases and sought Davy to be superintendent of experiments; at the same time he offered Davy the opportunity to continue his studies to become a surgeon. Although Beddoes was sympathetic to science, he seemed not to apply much acumen to his own experimenting. His sister-in-law reported that "one of his hobbies was to convey sows into invalids' bedrooms that they might inhale the breath of animals"—an idea that Beddoes was acting upon simply because he assumed it to be correct.

Joseph Priestley, discoverer of oxygen, was also impressed with Davy's first work. Priestley was then living in Northumberland, Pennsylvania, a refugee from intolerant views toward him. A political liberal, he had, in England, aroused the enmity of some of his neighbors with the political tracts he wrote. In one of his chemical pamphlets, he wrote: "Mr. H. Davy's Essays . . . have impressed me with a high opinion of his philosophical acumen. His ideas were to me new, and very striking, but they are of too great consequence to be decided upon hastily." Priestley's son later studied for a short time with youthful Davy.

One of Davy's first scientific accomplishments prior to going to Dr. Beddoes' establishment was to show the mistake in the view held by an American physician, a professor at Columbia College in New York, that nitrous oxide was a cause of disease. The scientist had publicized his ideas, among other ways, with a long poem sent to Dr. Beddoes, then at Oxford University. Davy wrote: "Wounds were exposed to its action; the bodies of animals were immersed in it without injury; and I breathed it, mingled in small quantities with common air, without any remarkable effects."

At Dr. Beddoes' Institute, Davy searched for a way to obtain large quantities of the gas nitrous oxide. He succeeded in making enough and immediately used it on himself. He recorded his sensation as "analagous to gentle pressure on all the muscles, attended by a highly pleasurable thrilling, particularly in the chest and the extremities. The objects around me became dazzling and my hearing more acute. Towards the last inspirations, the thrilling increased, the sense of muscular power became greater, and at last an irresistible propensity to action was indulged in." Dr. Beddoes thought the gas would be useful for paralytics, but Davy saw other possibilities. He inhaled some during the pain accompanying the growth of a wisdom tooth and wrote: "On the day when the inflammation was most troublesome, I breathed three large doses of nitrous oxide. The pain always diminished after the first four or five inspirations . . ." He concluded that "it may probably be used with advantage during surgical operations . . ." But nitrous oxide, laughing gas, was not to be used as an anaesthetic for another half century.

Davy next made the mistake of attempting to breathe other gases. He passed steam over charcoal to make water gas, a mixture of carbon dioxide and carbon monoxide, and wrote: "After the second inspiration, I lost all power of perceiving external things, and had no distinct sensation except a terrible oppression on the chest. During the third expiration, this feeling disappeared, I seemed sinking into annihilation, and had just power enough to drop the mouthpiece from my unclosed lips . . ." He concluded: "There is every reason to believe, that if I had taken four or five inspirations instead of three, they would have destroyed life immediately without producing any painful sensation." Later, French chemist Berthollet's son experimented similarly with carbonic acid, attempting to note all the successive feelings he experienced, and died.

Davy's work with laughing gas brought him popularity and an exciting new position. His poet friends Samuel Taylor Coleridge and Robert Southey unwittingly helped promote the use of laughing gas, and likewise did most participants at gatherings where the gas was inhaled for the gayety and hilarity provoked. In 1801, at the age of twenty-three, Davy was installed as assistant lecturer at the Royal Institution in London, an educational and research establishment founded in 1799 by Benjamin Thompson, Count Rumford.

Davy was not an instant success as a lecturer. Davy's official biographer records, "there was a smirk on his countenance, and a pertness in his manner, which, although arising from the per-

fect simplicity of his mind, were considered as indicating an un-
becoming confidence." But Davy was flexible, intelligent, and
quick to adapt to London society, its pretensions and policies.
About a month after he had come to the Royal Institution, he
joined a group of twenty-five men who called themselves the
Tepidarian Society because only tea was allowed at their meetings. The Tepidarians became ardent supporters of Davy, obtaining audiences for him before he became a sought-after lecturer. Then too, less than six months after he had come, the
full professor at the Royal Institution resigned because of ill
health and Davy was promoted to Lecturer.

Throughout 1801, Davy's talks were desultory. His success as
a lecturer began with one he gave January 21, 1802, presenting
what could be called twentieth-century ideas about the value
and power of science. He began to receive congratulations and
fan letters at the end of talks. His fame spread; soirees were
considered incomplete without the presence of Davy. Tickets
to his lectures became difficult to obtain. Once when he was ill,
the Royal Institution posted hourly bulletins about his health.
He adapted to the success by assuming "the garb and airs of a
man of fashion."

His popularity did not prevent accomplishments in science.
Although his laboratory procedure was not elegant and orderly
and he had a careless and sloppy style of work, he became one
of the founders of electrochemistry. He showed the true com-
ponents of water; discovered the chemical elements sodium,
potassium, calcium, barium, magnesium, and strontium; established chlorine as a chemical element; and found that hydrochloric acid was free of oxygen. In applied science, he invented a
miner's safety lamp, a device still in use today, competing successfully with more recently developed safety apparatus.

With his seemingly perverse talent for attracting adverse criticisms, he was not altogether inept in human relations. When
thirty-four years old, he married a rich widow and immediately
he was accused of seeking wealth, although Lady Davy could
have been accused of seeking the prestige that went with the
Davy name. To Davy's credit, he employed young Michael Faraday when Faraday was without much formal education.

Davy received a medal and three thousand francs from the

government of France while England and France were at war. He said: "Some people say I ought not to accept this prize, and there have been foolish paragraphs in the papers to that effect. But if two countries are at war, the men of science are not. That would indeed be a civil war of the worst description. We should rather through the instrumentality of science soften the asperities of national hostility."

He was one of the first to establish and demonstrate the international brotherhood of science. For one and a half years, he and Lady Davy, accompanied by Michael Faraday, toured Europe and were entertained by the men of science. Even though France and England were at war, he had the permission of the French government for his visit.

In outward appearances, Davy had made a great self-improvement. He was a country boy, minus family connections, who became well-to-do. His self-improvement is also obvious in scientific matters. Without much formal education, he joined the ranks of great scientists. One of his outstanding works not only illustrates a self-improvement leading to success, but also is a good example of the method of multiple working hypotheses.

At the turn of the nineteenth century, Davy heard of Italian scientist Volta's discovery of the chemical battery, and Davy, as well as others, began to experiment with this new tool. Nicholson and Carlisle showed that water could be decomposed into hydrogen and oxygen with the help of the battery current. Others found that not only the two gases but also an acid and a base were formed.

Speculative answers to the problem of acid and base formation in the electrical breakdown of water were plentiful. Luigi Brugnatelli said that electricity was a new acid. Another Italian, Pacchiani, said hydrochloric acid could be obtained from so-called pure water. Johann Ritter of Germany claimed that water was not a compound.

While still at the Pneumatic Institution, Davy had convinced himself that impurities of various kinds were responsible for the acid and base formation. He believed the acid, hydrochloric, came from the animal or vegetable matter used to connect the glass tubes of water undergoing decomposition. When he replaced this moist bladder with one of washed cotton, he did not detect the acid. In every case where much base was formed, the glass at the point of contact with the wire looked eroded. When Davy had substituted an agate cup for the glass one, base was not obtained.

When continuing this research at the Royal Institution, Davy already had the cause isolated to his own satisfaction. He used cups of agate boiled for several hours in distilled water. Instead of glass he used a piece of very white and transparent asbestos, also boiled for several hours in distilled water. Thus, every suspected cause of the acid and base formation was removed. The electrical action proceeded for forty-eight hours. At the end, the water in the cup holding the positive electrode gave indications of hydrochloric acid and that holding the negative electrode showed soda. The result was contrary to Davy's first hypothesis and preliminary work. Being convinced of the veracity of his idea, he accordingly wondered whether the agate could still contain some salty water. The experiment was repeated a second, third, and fourth time, and each time the amount of acid and base formed was reduced. Four additional repetitions convinced Davy that although agate was one of the sources sought for, it was not the only one. A scientist of lesser stature would have accepted this dilemma by retiring to an easier problem or citing the anomaly as inconsequential. But Davy was a great scientist, ever correcting his own views when the facts would not support them.

Since the apparatus contained some sources of the acid and base, he used a pure gold vessel—he could then afford it. The experiment was repeated, and this time the acid formed was traced to a reaction with nitrogen of the air while the base was traced to the water. In his quest for complete understanding, he next used ultra-pure water; he made it by evaporating water in a silver vessel. The purified water was subject to hours of electrical current. The result was satisfying; the cause of the base was finally traced. In order to show definitely that nitrogen of the air was partially responsible for the acid formation, he exhausted the air about his equipment; and after eighteen hours of electrical action, only a small amount of acid was produced. Davy sought even more favorable evidence. He exhausted the air around his apparatus and filled it with hydrogen. He made a second exhaustion and again introduced carefully prepared hydrogen. Electricity was continued for twenty-four hours, and at the end, not a trace of acid or base could be found.

The 1966 Nobel-prize winner in physics, American Robert S. Mulliken, said it this way: "The scientist must develop enormous tolerance in seeking for ideas which may please nature, and enormous patience, self-restraint, and humility when his ideas over and over again are rejected by nature before he arrives at one to please her."

SELECTED REFERENCES

Crowther, J. G. *Men of Science*. New York: Norton, 1936.

Goran, M. "The Roots of Scientific Method," *School Science and Mathematics,* 53 (February, 1963), 115–18; reprinted in E. W. Courtney, Ed., *Applied Research in Education*. Paterson, N.J.: Littlefield Adams, 1965.

———. "The Next Step in Science Education," *Improving College and University Teaching*. (To be published, Autumn, 1971.)

Hartley, H. *Humphry Davy*. Camden, N.J.: Thomas Nelson, 1967.

Paris, J. A. *The Life of Sir Humphry Davy*. London: Henry Colburn and Richard Bentley, 1831.

Williams, L. P. "Humphry Davy," *Scientific American,* 202 (June, 1960), 106–16.

INDEX

149

DATE DUE

GAYLORD			PRINTED IN U.S.A.